metaLABproject

The *metaLABprojects* series provides a platform for emerging currents of experimental scholarship, documenting key moments in the history of networked culture, and promoting critical thinking about the future of institutions of learning. The volumes' eclectic, improvisatory, idea-driven style advances the proposition that design is not merely ornamental, but a means of inquiry in its own right. Accessibly priced and provocatively designed, the series invites readers to take part in reimagining print-based scholarship for the digital age. www.metalab.harvard.edu

Todd Presner
David Shepard
Yoh Kawano

‡‡‡‡‡‡‡‡‡‡‡‡‡‡‡‡‡‡‡‡‡‡‡‡‡‡‡‡‡‡

HyperCities
Thick Mapping in the Digital Humanities

metaLABprojects

Harvard University Press
Cambridge, Massachusetts, and London, England
2014

Printed in China

Cataloging-in-Publication Data available from the Library of Congress

Graphic Design:
xycomm (Milan)
Gennaro Cestrone
Stefano Cremisini
Francesca Farro
Edda Bracchi

Table of Contents

Preface

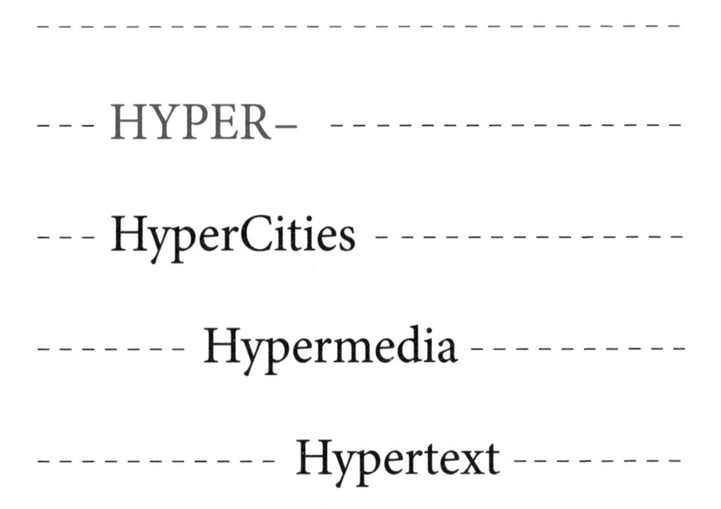

(TP) The prefix "hyper" refers to multiplicity, abundance, and heterogeneity. A hypertext is more than a written text; hypermedia is more than a single medium; and HyperCities are more than the physical spaces of cities. "Hyper" adds to, extends, and proliferates: many texts and media, many extensions to and contestations of the historical record, and an infinitely open field of possibilities for participation. A HyperCity is a real city overlaid with thick information networks that not only catalyze the present but also go back in time to document the past and go forward to project future possibilities. HyperCities are always under construction.

This is a book about humanities explorations of digital mapping. That's strange, you say. **What do the humanities have to do with map-making? And why write a book about something digital?** To address the first question: the book brings together the methods, content, and values of the humanities to create and re-create representations of places, what we are calling "thick mapping." The methods are design-centered, critical interrogations and interpretations of the multiplicity, heterogeneity, and ambiguity of the cultural record of places. It is not a book about "maps" per se but about exploring, participating, and listening, something that transforms our conception of mapping into a practice of ethics. At the core of this book is an idea about the possibility of expanding participation and of the value of knowledge in the service of the public good.

But why write a book? A book allows for the choreography of an argument. While you can certainly start at any point in the book or skip around at will, if you read the pages sequentially, you will be brought through the argument as a choreographed experience, much like watching a dance ensemble perform a piece from start to finish. The authors imagine the book as a stage upon which a performance takes place, one in which many voices and movements come together in time. In this sense, the book is not a simple linear narrative; rather, it is a staging of various kinds of narratives and media in a spatial arrangement that is cumulative and recursive. In other words, the book represents a practice of mapping, an emplotment of many narratives and voices at once.

It starts with a series of essayistic pieces that define the key terms (HyperCities, thick mapping, digital humanities) before turning to a conceptual and technological genealogy of the HyperCities project. In places throughout the book, the narratives run parallel to a series of critical code studies.

We examine modules of code in order to expose the undergirding assumptions and structuring principles built into the HyperCities project at the level of the code itself. From there, the first project windows open up. These windows are written in the voices of the authors, designers, and cartographers themselves. After that, the book presents a media archaeology of Google Earth and turns to several cultural-historical investigations of map projections and the political epistemologies of mapping. This chapter tarries with the profound contradictions and fraught dialectic at the heart of the HyperCities project and digital cultural mapping more generally. The final chapter focuses on two world-historical events—the 2011 "Arab Spring" in Egypt and Libya, and the Japanese earthquake and tsunami that same year—through the lens of social media mapping. The investigation moves from aggregations of data and high-altitude visualizations to the singularity of the human voice that personalizes and punctures any abstracted totality. The narrative changes from theoretical and conceptual explorations of "events" to first-person accounts, photographic documents, and fragmentary, 140-character long Twitter streams. The geolocated tweets of a single eyewitness, Tranquil Dragon, are mapped onto the space of the page and can be "played" like a miniature flip-book.

Throughout the book, we can also make use of certain rhetorical devices, tropes, and figures such as synecdoche, ekphrasis, and aphorism. We consciously move between verbal and visual modes, while writing in several different registers, sometimes even on the same page, in order to denaturalize expectations of a single, objective voice. At the same time, the authors assume multiple relationships to and perspectives on the material, the events, and the objects of study. We take up these positions from within the act of writing and composing, as a kind of middle voice, in which neither active nor passive speech predominates. This is also driven by an awareness of our (sometimes) precarious ethical stance and embodied relationship vis-à-vis the material under consideration. Sometimes the view is "zoomed out" from a perspective of abstracted data, and sometimes the view is profoundly "zoomed in," such as Kawano's experience of going inside the radiation zone around the Fukushima nuclear power plant. This is a view that is no longer that of a safe spectator but of a body-at-risk.

More pointedly, the book is an experiment in remediation, in which certain expectations of genre and narrative are consciously flouted. "Scholarly" writing exists side-by-side with code, which exists side-by-side with community generated content and voices from the public-at-large through social media feeds. As the book progresses, the long-form narratives break down into smaller and smaller pieces such as data streams, tweets, and images to reveal different perspectives—distant and close—on the mapping of events. The last chapter literalizes Walter Benjamin's famous injunction that "history breaks down into images, not into histories [or stories, or narratives]."[1] Benjamin was playing with the German term for history (*Geschichte*), which can mean an event (in the sense of *das Geschehen*, something that happened) as well as "history" or "story" in the sense of a narrative account of events and occurrences. For Benjamin, there is no homology between that which happened and the narrative means to present the past; instead, there are only transitory, fragile, and dialectical images, which form montages of interplay and constellations of tension between that which was and the now-time of a given present.

Altogether, you are looking at a book that was collaboratively conceived, authored, and choreographed at every stage of its development and design. The primary authors were Todd Presner (Minion Pro), David Shepard (OCRB), and Yoh Kawano (Neuzeit S). Not unlike Mark Z. Danielewski's experimental novel *House of Leaves*, the different fonts within the book indicate changes in narrative voices. The book also features contributions written and designed by project leaders that open up as "windows" onto the HyperCities idea and digital cultural mapping more generally. There are four such windows in the book: a thickly layered "ghost map" of downtown Los Angeles by Philip Ethington; a cultural history of Historic Filipinotown (HiFi) in Los Angeles by Mike Blockstein and Reanne Estrada, told through multiple mappings of this neighborhood using personal stories, art installations, and video testimonies; a four-dimensional geotemporal argument and virtual worlds exploration of the Roman Forum in the ancient world by Diane Favro and Chris Johanson; and, finally, a day-by-day and hour-by-hour mapping of the Tehran election protests of 2009 by Xárene Eskandar using social media.

In selecting these four windows, we wanted to showcase some of the variety and texture of mapping projects in the digital humanities. They are each told through the voices of the primary project cartographers and exemplify innovative approaches to creating and deepening public knowledge. In the case of the HiFi project, for example, many of the voices are those of immigrant youth living in Historic Filipinotown, collaborating with a class at UCLA, the Pilipino Workers Center of Southern California, and a social enterprise called Public Matters to create a cultural record and to preserve community memories of this neighborhood. In the case of Tehran, the maps are attempts to curate the voices and images of protest in the face of powers determined to suppress and erase them.

All these voices and maps come together not in a hypertext (which continuously changes) but a hyperbook about HyperCities, a "build" of a digital platform in the form of a book. It is a polyvocal, multilevel form inspired by and constructed with digital technology, manifested in print because a printed book can assemble these multitudes into a snapshot before they separate again. Or not a snapshot, perhaps, but a timed exposure: many of these projects and ideas took years to assemble, and we hope they will persist. This record marks them as enduring but acknowledges they may change or disappear. It is an act of remediation and transmission, as much as an act of preservation, of the ephemerality of the digital.

Lexicon

HyperCities

(TP) Imagine a digital narrative crisscrossing place and time, starting with the date and location of your birth. The narrative grows, fragments, and connects many places and times together, as your life unfolds and as you tell your story. Any event in your life can be geo- and time-located, and each event connects with innumerable other events in your life and the lives of others. Everyone's life story intersects with countless others at every moment, creating ever denser webs that document the complexity of the human experience. Every story matters, every voice can be heard, every event—no matter how big or how small — can be captured.

What if the World Wide Web was composed of lives, not pages? Imagine a search and discovery tool for this web in which you traveled through time and space, rather than inputting keywords in an empty white box. Imagine browsing this web by deciding where and when you wanted to go rather than what you needed to find. What if you could "drill down" at a given place over the last ten seconds, the last ten minutes, the last ten months, the last ten years, or the last ten centuries? Among other things you might find historical maps, architectural reconstructions, personal stories, pictures, documents, and other sources that bring alive the texture and vibrancy of past places. What if you could search any place in the world on a single day, and what if these ever-expanding, ever-thicker webs of events, documents, and lives were keyed to physical locations in built spaces? What if there was a way to let the ghosts return, the memories come to life, and the silenced voices be heard? Would it be a vision of digital democracy or a dystopia of total knowledge and control?

Alas, HyperCities doesn't exist. It's an idea, largely unrealized, perhaps impossible. This book is an exploration of attempts to imagine, build, and tarry with something called HyperCities. It documents a series of "laboratory" projects inspired by this idea. If you are looking for a finished software platform called HyperCities, let us be clear: it doesn't exist. There is no fully functional, released piece of software, only a series of prototypes, experiments, code modules, and projects that more or less work. This book is a synthesis of the HyperCities idea, composed in historical, conceptual, and speculative modes. It tells the dialectical story of an idea—through many voices and with many authors—in the burgeoning field called "thick mapping in the digital humanities."

Semantically, the term "HyperCities" accords with "hypertext" and "hypermedia," coinages by the visionary media theorist Theodor Nelson.[2] In a seminal essay of 1965, "A File Structure for the Complex, the Changing, and the Indeterminate," Nelson sketched out an open-ended, non-linear system for organizing, interlinking, and accessing information. A hypertextual structure, in his articulation, cannot be reduced to a single medium (such as print) and could grow and change as new information was added to the system. The term "hypertext" thus refers to "a body of written or pictorial material interconnected in such a complex way that it could not conveniently be presented or represented on paper . . . Such a system could grow indefinitely, gradually including more and more of the world's written knowledge" (144). Anticipating the World Wide Web by nearly twenty years, Nelson called his invention the "Evolutionary List File," an interconnected, interlinked, hypermedia information system that could grow and proliferate as users added new material. As an open-ended authoring, curatorial, and annotation environment, HyperCities is founded on an analogous logic.

Within the disciplines of geography and urban planning, "hypercities" sometimes refer to densely populated cities with more than twenty million inhabitants, so-called "megacities." While this definition is relevant for investigating urbanism and population dynamics, the HyperCities idea focuses on the past, present, and future of cities. In this respect, HyperCities are much larger than twenty million inhabitants, since they might embody the entire (largely erased or absent) history of a city—potentially every life, every structure, every street, every voice over time. Of course, it would be absurd to posit the resuscitation of the dead or pursue a naïve positivism (that the past can be fully recuperated or represented by the employment of technology); instead, HyperCities is about the possibility of telling stories, of narrating places, and of producing new configurations of knowledge in which every past, present, and future is a place. In this sense, mapping history is about curating places, conjuring and caring for ghosts.

Thick Mapping

Mapping is not a one-time thing, and maps are not stable objects that reference, reflect, or correspond to an external reality. Mapping is a verb and bespeaks an on-going process of picturing, narrating, symbolizing, contesting, re-picturing, re-narrating, re-symbolizing, erasing, and re-inscribing a set of relations. On its most fundamental level, a map is a graphical representation of a set of relations. Maps are visual arguments and stories; they make claims and harbor ideals, hopes, desires, biases, prejudices, and violences. They are always relational, in dialogue or in contact with someone or something. They may or may not attempt to reference, reflect, or represent an "external reality" (however one defines that), but they are fundamentally propositions, suffused with world-views, structuring epistemologies, and ways of seeing. Maps are representations of a world, which reference other such representations. When we georeference historical maps, we are not "correcting" them or making them "accurate"; instead, we are keying one representation to another representation (not to reality).

The history of cartography indicates a clear recognition of the material substance or media of the map. The very terms "map" and "chart" derive from their materiality: the Latin word *carta* denotes a formal document on paper or parchment, while the term *mappa* indicates cloth.[3] Prior to the printing of maps in the late fifteenth century, maps were often drawn on parchment or cloth or etched in wood, metal, or stone. In Renaissance Europe, the circumnavigation of the world and the production of accurate projections for empirical exploration went hand-in-hand with the engraving of world maps and the production of celestial and terrestrial globes. With the development of the printing press and the scientific revolution in the Age of the Enlightenment, mapping began to assume a central role in developing accurate statistical methods (such as the census) and the proliferation of mappable data, both of which played a critical role in the expansion of the European nation-state and the colonial conquest of the "unknown" world. Not until the late nineteenth and early twentieth centuries did non-print technologies (particularly, aerial photography and film) play a role in producing maps of the world. This would be taken to a new level with the deployment of remote sensing Global Positioning Satellites (GPS), allowing accurate determinations and targeting of any point on earth according to latitude, longitude, altitude, and time. With the development of the first computational tools for producing digital maps and analyzing troves of geo-data in the 1970s, the material history of mapping entered a new chapter: mapping was transmogrified into computational processes and Geographic Information Systems (GIS). Today, web-based mapping applications such as Google Earth, OpenStreetMap, and WorldMap have brought the analytic tools of GIS to the general public and are changing the way people create, visualize, interpret, and access geographic information.

Digital mapping offers a significant break in the history of cartography precisely because it fundamentally changes the materiality and media of mapping. Unlike artifactual maps on paper, cloth, or parchment, digital maps are extensible, mobile, and networked. As such, new data about location (ranging from traffic reports, crime statistics, voting patterns, and housing prices to user-generated routes, historical photographs, and personal stories) can be instantly added from a range of web-enabled devices. But until recently, these tools have primarily served utilitarian purposes (like driving directions), as well as, more ominously, micro-level surveillance and population monitoring. Following the 2012 presidential race, massive amounts of GIS data, strategically keyed to and targeted at the granularity of a given household, were touted as critical determinants for Obama's reelection. And following the revelations of the scope of the NSA's dataveillance programs, we now know that we live in a world in which everything and everyone can be watched, monitored, tracked, and mapped. "Thick mapping" has an underbelly of unmitigated paranoia and unchecked control.

On its most basic level, "thick mapping" refers to the processes of collecting, aggregating, and visualizing ever more layers of geographic or place-specific data. Thick maps are sometimes called "deep maps" because they embody temporal and historical dynamics through a multiplicity of layered narratives, sources, and even representational practices.[4] But "thickness," as we are using the term here, is not quite tantamount to "depth." Depth models, of course, abound in the history of modernism: Freud imagines psychoanalysis as an archaeological enterprise, likened to unearthing ancient Rome, in which the latent desires of the subject can be probed ever more deeply; hermeneutical models in the sciences and certain historical disciplines imagine their methods as excavations of hidden processes and meanings; the aesthetic forms of modernism—the black square on a black canvas, the glass box, but also the montage form—stemmed from a world in which deep, total, and utopian "solutions" were still imaginable and possible. Postmodernism, on the other hand, is supposedly all about surface, the infinite play of signifiers, the total loss of historicity, and the schizophrenic subject called to cognitively map the world in order to somehow brook resistance to the leveling effects of capitalism. It privileges categories of spatiality precisely because the mutation in the global spaces of multinational capital requires the development of new perceptual habits to find orientation, develop agency, and map this space.[5]

HyperCities draws from both modernism and postmodernism: it is inspired by a depth model rooted in the idea of archaeological coring and can be seen as a response to the crisis of historicity. And yet it is infinitely extensible and rhizomatic in practice, simultaneously moving vertically and horizontally, down and across. Intertextual play exists side-by-side with historical layers of meaning-making; practices of cognitive mapping are both global and local but never simply mimetic, as if a stable external reality can be reliably and definitively mapped.

Instead of positing another depth model or yet another celebration of postmodern hyperspace, the HyperCities project strives for "thickness." Thickness means extensibility and polyvocality: diachronic and synchronic, temporally layered, and polyvalent ways of authoring, knowing, and making meaning. Not unlike the notion of "thick description" made famous by anthropologist Clifford Geertz, thickness connotes a kind of cultural analysis trained on the political, economic, linguistic, social, and other stratificatory and contextual realities in which human beings act and create.[6] By eschewing any kind of universalism, it is a kind of analysis that is intrinsically incomplete, always under contestation, and never reaching any kind of final, underlying truth. Thick mappings, like thick descriptions, emphasize context and meaning-making through a combination of micro and macro analyses that foster a multiplicity of interpretations rather than simply reporting facts or considering maps as somehow given, objective, or complete.

Thick maps are conjoined with stories, and stories are conjoined with maps, such that ever more complex contexts for meaning are created. As such, thick maps are never finished and meanings are never definitive. They are infinitely extensible and participatory, open to the unknown and to futures that have not yet come. And perhaps most importantly, thick maps betray their conditions of possibility, their authorship and contingency, without naturalizing or imposing a singular world-view. In essence, thick maps give rise to forms of counter-mapping, alternative maps, multiple voices, and on-going contestations. Thick maps are not simply "more data" on maps, but interrogations of the very possibility of data, mapping, and cartographic representational practices. In this sense, "thickness" arises from the never-ending friction between maps and counter-maps, constructions and deconstructions, mappings and counter-mappings.

Digital Humanities

The conjunction of "digital" and "humanities" raises fundamental questions for documenting and analyzing the cultural record of humankind. "Digital" is a shorthand term that connotes the domain of the computational governed by binary numeric form and the electronic technologies that operate according to this logic. The Internet and the World Wide Web are, of course, digital technologies but the digital refers, more broadly, to any computational or algorithmic procedure to encode, present, distribute, and analyze data. This logic is, ostensibly, antithetical to the "humanities" which are, at least traditionally, the domain of the arts, philosophy, literature, and culture more generally. The humanities are characterized by creative energies and critical practices that relish ambiguity, subjectivity, and interpretation. They cannot be reduced to ones and zeros.

But over the past decade, the methods, media, and materiality of humanities research have undergone dramatic change, with massive new possibilities emerging for authorship, creative design, meaning-making, data curation, interaction, and dissemination of scholarship. The world of print culture has not vanished, but it has become transformed in fundamental ways and supplemented by new technologies that allow researchers to ask entirely new questions about the cultural record, at a scale that requires computation. As such, the humanities have developed new research methods through their encounter with the computational sciences, not only creating large and complex cultural datasets for analysis but also fostering humanistic approaches to algorithmic thought, which interrogate the governing assumptions built into technologies, data, and computational practices themselves. "Digital Humanities" is an emerging field that explores the deeply productive tension and precarious linkage between computational practices and humanities scholarship. The HyperCities project is a product of this linkage.

This is why HyperCities is not primarily a "technological" or "computational" problem but foremost a "humanities" problem, namely one of memory, narrative, archival practices, knowledge design, and, finally, ethics. The Digital Humanities for which I am arguing is not simply about computational processing of data but about the design of something new, an "insertion"—as Hannah Arendt might say—of a new potentiality, of a future that remains open to possibilities, even new worlds. We thus begin by inserting ourselves into the world.

The Humanities in the Digital Humanities

"The street conducts the flâneur into a vanished time. For him, every street is precipitous. It leads downward ... into a past that can be all the more spellbinding because it is not his own."

– Walter Benjamin, *The Arcades Project*, 1928-40

(TP) The figure of the flâneur—the so-called "man of the crowd"—was made popular in the mid-nineteenth century by the likes of Edgar Allan Poe and Charles Baudelaire in short stories and poetry that portrayed the modernity of great cities like London and Paris. The flâneur (almost always a man of leisure, a dandy) strolled along the bustling streets, under the gas lamps, observing the metropolis as an urban spectator and occupying a liminal zone of privilege and transgression. But for Walter Benjamin, the German-Jewish cultural critic and philosopher writing his magnum opus, *The Arcades Project*, a book on nineteenth-century Paris, the flâneur was not simply someone who walked the streets of the modern metropolis and disappeared into the swarm of crowds; rather, it was someone who was a time-traveler. As the flâneur walked along the streets, he was conducted downward in time. What a striking idea: that the physical topography of the street could lead you back to a time that had vanished, to a time that was not even your own. How could this be? Is it really the street, or might it be a kind of sensibility or openness to apprehending, listening to, and, ultimately, caring about and caring for the past? In other words, maybe the past is always there—quiet, muted, faded, hidden—and it is the task of the flâneur to enable it to speak, to make it come alive and come to light, and, thereby, resonate with the present. In this sense, the past must be conjured, awakened, and cared for.

Legibility / Recognizability

(TP) Driving around Los Angeles, I use Google Maps almost every day. The streets are yellow, the buildings are white, the parks are green, and the waterways are blue. Street names are written in black Arial font with a white glow. The maps are clean and move effortlessly with my finger. They seem infinitely draggable and zoomable. The browser window on my laptop or phone is hardly a limitation; these maps are slippy beyond all enframing. And they grow thicker, as multiple data layers are toggled on or off: traffic, satellite imagery, photos, webcams, and more. I get driving directions and check for traffic using Google's real-time data. Don't take the 405 at this time of day. Clicking on the little orange person, I am taken out of my world of stereoscopic vision to one constituted by nine camera eyes and stitched together to form a panoramic digital bubble that lets me see streets, interiors, and even oceans 360 degrees horizontally and 290 degrees vertically. I start to see differently, as if I'm flying above the world, zooming in and zooming out at will, in a multi-perspectival digital bubble. What does it mean that this seamless panning and zooming has become (almost) naturalized, that it has become how I see and experience the world, or how I *want* to see and experience my world?

I turn off Google Maps and start to drive. I wonder: what would it mean to drive downward, into the buried pasts that persist somewhere—in the imagination, in the archive, in the memories of others, in the traces of places long gone and lost? Who used to live here? What used to be there? What's buried under this freeway, under this skyscraper, beneath these overpasses? What has vanished imperceptibly from the surface of the earth? What voices and ghosts haunt, however imperceptibly, these concrete landscapes? Why do I care? It's a past which is, ostensibly, not my own. I don't recognize it. It's not mine. Time is out of joint.

Berlin, November 1995. Six years earlier, I had watched the Berlin Wall tumble on an eighteen-inch television screen mounted above the blackboard on the wall of my high school history classroom. Now, I was living in Berlin, near Mitte, a region in the east that had once been the center of Berlin's eastern-European Jewish population. The outer walls of buildings still had bullet holes from street fighting during the last months of the Second World War. Cranes dotted the city skyline, and brightly colored pipes, pink and green, ran along and over almost every street. Water was constantly being pumped out of the ground. Because the entire city is built on sand, the water table is only a few meters deep. Friedrich Schinkel, arguably Berlin's greatest architect, knew this when he decided to build the Altes Museum, a monumental Greco-Roman building, on a massive pile of logs to prevent it from sinking. Hitler's architect, Albert Speer, was also aware of the problem and engineered a massive pylon structure to hold up Hitler's Triumphal Arch, intended to mark the new southerly entrance into the city and celebrate the thousand-year empire. The pylon is a solid cement cylinder that measures 14 meters (1)

high, 21 meters in diameter, and extends 18 meters underground; it weighs 12,650 tons and exerts a pressure of 72 tons per square foot. It's still there today (you can see it in Google Earth), and it can't be destroyed because it is too close to residential apartments. (Fig. 1) And so it remains in a middle-class neighborhood in southern Berlin, as a strangely protected monument to a megalomaniacal dream. The triumphal arch, of course, was never built.

The Wall divided the city for twenty-eight years, or, more precisely, encircled West Berlin and turned it into a little island in the middle of East Germany. Subway stops located in the east turned into "ghost stations," as underground trains originating in the west could no longer stop in East Berlin. Above ground, long-distance railway lines that once connected Berlin to all of Europe sat idle, cut off by the Wall. Over decades, trees grew between the railway tracks; stations languished in ruin, bombed out by the aerial war and left to the forces of nature as the Cold War raged on. I became obsessed with ruins, perhaps because I wasn't used to living among them. For Berliners, they were as unremarkable as any other feature of the natural or built environment. Berlin is a very flat city, but there are two mountains in the neighborhood of Friedrichshain, both covered with grass and even hiking trails. They are man-made mountains of debris, the remains of two million cubic meters of destroyed buildings that were firebombed in the last years of the War.

Other ruins, like those of Berlin's proudest railway station, the Anhalter Bahnhof, and the railway tracks leading away from it, were simply fenced off, stranded in another time and left to the future to figure out what to do with them. They were finally disposed of in 2008, more than half a century after the last train departed the railway station. Only a small fragment of the entrance portal to this once-great railway sta-

tion remains, sitting idly beside a busy street and soccer park, which exists on the empty ground where the station once stood. (Fig. 2) The past was recognizable, but only just barely.

(2)

Time (the simultaneity of the non-simultaneous) / Space (the contiguity of the non-contiguous)

(TP) Great cities like Berlin are almost inconceivably complex and multilayered. Over its nearly eight centuries, Berlin emerged from a backwater mercantile town built on sand to become the capital of a unified Germany under Bismarck and the site of Hitler's dream for a world-dominant Germania. It was devastated by the Thirty Years War, occupied by Napoleon in 1805, rebuilt numerous times throughout the eighteenth and nineteenth centuries, destroyed by the aerial bombardment campaigns of World War II, divided by the Berlin Wall for twenty-eight years, and hastily put back together again in 1990. Poised on the border between Western and Eastern Europe, this cosmopolitan city has variously welcomed and persecuted its minorities: Huguenots, Jews, Poles, Russians, Turks, and others. It doubled in size in less than a quarter of a century between 1890 and 1925, reaching a size of four million people, and making it one of the largest cities in the world at the time. Another quarter of a century later, it lost almost half of its population and nearly all of its Jewish population in World War II and the Holocaust. Berlin, like other great cities, is comprised of densely layered architectural, social, political, and cultural palimpsests. I began to wonder: how can one make sense of this complexity, the many pasts, hopes, fears, and desires built and buried in these urban landscapes? How could one detect, mark, or hear the voices that had once traveled in these places, many of which are now just ruins? What did it mean to walk, as a flâneur, along these streets of amnesia? I began making mental maps of vanished times.

In the mid-1990s, I was hardly the only one to engage in such a critical, cartographic practice of memory and media. It was at this time that I encountered the Berlin interventions of the ART+COM group, particularly their multimedia project "The Invisible Shape of Things Past" (1995–2007).[7] This interactive mapping and visualization project, focused on Berlin's Leipzigerplatz, placed historic film sequences as three-dimensional spatial objects on top of time maps. The result was a series of complex time-space perspectives and navigational interfaces to probe the layered histories of Berlin, some of which were still visible but most of which had slipped into oblivion. Like "The Invisible Shape of Things Past," the maps I began to create as a prototype for the

HyperCities project were strangely dense—like stacks of images overlaid one on top of the other—that did not quite line up and offered little in the way of clarity. That's because all of these pasts co-exist, in various degrees, depending on what remains legible and what we care to recognize in the spaces of the present. A church from the thirteenth century sits next to an imposing television tower, encircled by the sedimented remains of the medieval city wall and a moat, on which a railway line runs and crisscrosses the city that was once divided into two halves. The solid cement cylinder commissioned by Speer in 1937 was the foundation of a future that—thankfully—will never come to be; but given its ponderous size and solidity, the cylinder, ironically, may easily remain for a thousand years. These pasts are not simply part of some long-gone, chronological history but exist—simultaneously—in the spaces of the present. What's muted are the hopes and desires, anxieties and fears that were embodied in the ways these vanished pasts imagined the future. With few exceptions, all but the most strident voices are lost, all but the most indelible memories are erased—and even these are transient.

The space of the modern metropolis is indissociable from and profoundly shaped by the various media in which it is represented, whether through novels, photography, film, television, computer simulations, or geo-browsers like Google Earth. City films, such as those by Walter Ruttmann, Dziga Vertov, and Sergei Eisenstein, gave rise to a new phenomenology of the city, as space was fragmented into its constitutive parts and reassembled according to the perceptual contingencies of the film maker. As Benjamin explained with regard to film, the embodied experience of space and time changed with the advent of this new medium: "Our taverns and our metropolitan streets, our offices and furnished rooms, our railroad stations and our factories appeared to have us locked up hopelessly. Then came the film and burst this prison-world asunder by the dynamite of the tenth of a second, so that now, in the midst of its far-flung ruins and debris, we calmly and adventurously go traveling. With the close-up, space expands; with slow motion, movement is extended."[8] The embodied experiences of the nineteenth century flâneur were now displaced by the non-contiguous, disembodied representations of the city that can be seen or heard from a distance, at varying tempos, and even in a new sequence. Not only is space transformed by new media, allowing it to be apprehended and experienced in ways that were previously not possible, but the figure of the flâneur is also transformed, as spectatorship migrates to the screen and eventually the computer interface. Today, we are all digital dandies, and thus it is no coincidence that the figure of the flâneur has frequently informed discussions of the co-constitutive relationship between urban modernity and new media, whether photography, film, computer simulations, or other digital technologies.

In film, the realism of urban space is reconfigured and reordered through the principle of "the contiguity of the non-contiguous." In Ruttmann's *Berlin: Symphony of a Great City* (1928), objects, buildings, streetcars, train stations, factories, theaters, and entire neighborhoods of Berlin are remapped through cinema and placed side-by-side *as if* contiguous. The result is that the new media flâneur can experience the city in a way that was simply not possible for Poe's man of the crowd or Baudelaire's dandy, both of whom physically walked around the built space of the city. Now, a viewer of Ruttmann's film can travel through and experience Berlin synchronically: filmed over a year, a single, twenty-four day is presented in an hour of film, and scores of physically non-contiguous locations are rendered consecutive and contiguous—precisely the way a traditional flâneur could never perceive or experience the city. "What if . . ." is the foundational principle of navigation of the city in every age of new media.

HyperCities: A Very Brief History

2000: Web 1.0, the Readerly Web

(TP) The original idea for the digital mapping project that would eventually become "HyperCities" was a research initiative called "Berlin: Temporal Topographies" that was begun around 2000 at the Stanford Humanities Laboratory. Temporal topographies were just that: investigations of the historical and time-based dimensions of places. The concept refers to the inscription of time in space, the simultaneous presence of different "time-layers" in a given location. The initiative yielded an early website that sought to present a multiplicity of pathways through the city. The pathways were HTML web pages about aspects of Berlin's cultural history, illustrated with photographs and other digitized artifacts. The idea was to eventually have many authors who created pathways, some of which would intersect with one another, and others of which would veer off in new directions. While the project never matured beyond the initial website, it yielded an array of conceptual approaches that would become foundational for the burgeoning field of digital cultural mapping and the HyperCities project itself.

In retrospect, I would call this project Berlin 1.0, the readerly web, in which knowledge is presented to viewers who are asked to consume it but not alter its conditions, meaning-making strategies, and modes of production. The readerly web is ultimately just that: a web of pages to be read, a web governed by a singularity of meaning achieved through a common set of naturalized practices of interaction and consumption.

2004: Flash—Animating and Reanimating the Readerly Web

(TP) In 2004, I designed a flash-based, digital textbook called "Hypermedia Berlin" with the support of UCLA's Center for Digital Humanities. Tracing its historical genealogy back to "database" projects such as Ruttmann's city film and Benjamin's cultural history of nineteenth century Paris, the goal of Hypermedia Berlin was to construct a web-based platform for representing and studying the cultural, urban, and architectural history of a layered city space. While Benjamin attempted to create a montage text to investigate Paris, we were using the technologies of new media to imagine a

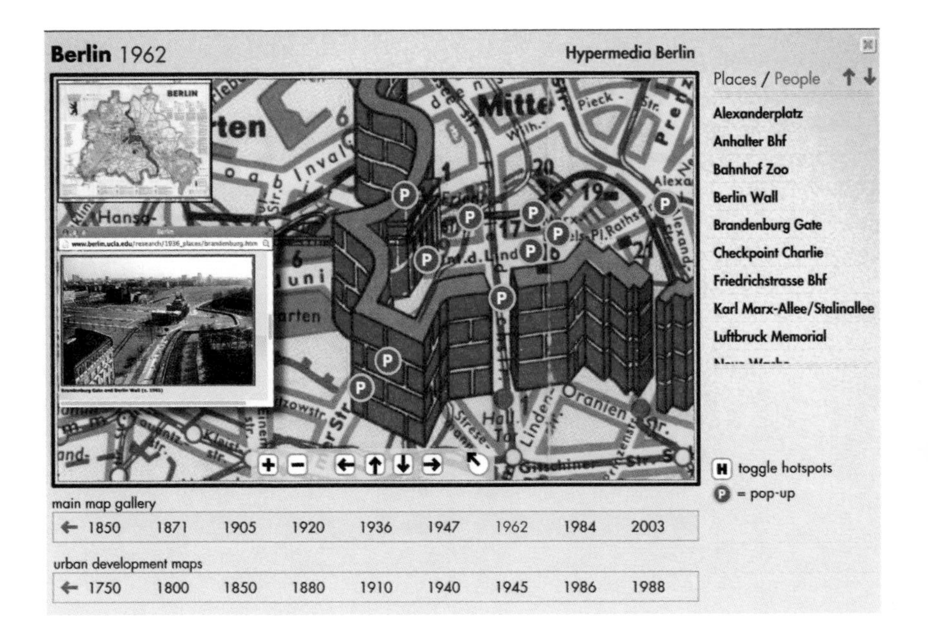

new study of city spaces and culture, something that Benjamin was ultimately unable to realize in the hand-written folios he made for *The Arcades Project*. (Fig. 3)

(3)

Deploying the organizing principle of "temporal topographies," my team created a digital textbook for exploring Berlin using a series of digitized historical maps that I manually georeferenced as the basis for experiencing the city's history from 1237, when the city was founded, up through the present day. But rather than taking chronology as the sole organizing principle, I wanted to foreground the uneven time-layers comprising Berlin's nearly 800-year history, the urban palimpsests, as it were, that form the present. "Hypermedia Berlin" was completed in 2004 and allowed students to navigate by both time and space through twenty-five interlinked maps of Berlin keyed to relevant "people" and "place"

http://www.berlin.ucla.edu/hypermedia

links. Students explored Berlin by zooming in and out of the maps, scrolling—in any order—through some 800 years of space and time, and clicking on various regions, neighborhoods, blocks, buildings, and streets.

Technically, the project was constructed in Flash, using HTML popup windows, and a modified version of "Zoomify," a software program that allows viewers to zoom in and out of particular locations of an image. A visitor might jump to the Berlin of 1811 or 1850 and then proceed, by place, through the Brandenburg Gate in 1962, 1936, and 1871. In each "time-layer," photographs, film clips, and popup "windows" detail aspects of the cultural and architectural history of the city at that moment, including extensive information about key people in Berlin's history. I taught the pilot version of "Hypermedia Berlin" at UCLA in the Spring of 2004 to a class of 65 undergraduate students. Applying the theory of the "simultaneity of the non-simultaneous" as an organizing principle, every lecture began at a place in present-day Berlin (such as Potsdamer Platz or the Olympic Stadium) and then proceeded to "drill down" into the historical layers impacted in these places. At any moment, students could stop and "travel" synchronically or move backward or forward in time to other city layers. The class seemed to elicit great curiosity, intellectual excitement, and not a small bit of confusion.

2005: Geo-temporal Search and the Writerly Web

(TP) When Google released its Map API in the summer of 2005, a small revolution occurred: anyone with basic programming skills could now integrate Google's world map and the accompanying satellite imagery into individual websites, create and mark up maps using this imagery, and even develop new software using the Google Maps application. An Application Programming Interface (API) allows programmers to build on, customize, and incorporate existing software code into their own applications. Quite suddenly, the world of Geographic Information Systems (GIS), a terrain that had been dominated for decades by significantly more abstruse, desktop applications such as ESRI's ArcMap and ArcGIS, had been opened up to the masses. Map mash-ups flourished almost overnight, as nearly everyone with any sort of geo-data began making maps—of their favorite restaurants, bike paths, and whale-watching spots. Geo and time mark-up became indispensable metadata fields for a vast array of web content, prompting Michael Jones, Google's Chief Technologist, to emend his company's mission "to geographically organize the world's information and make it universally accessible and useful."[9]

Built on the idea that every past is a place, HyperCities came to life as a digital research and educational platform for exploring, learning about, and interacting with the layered histories of city and global spaces. Developed though collaboration between UCLA, USC, CUNY, and numerous community-based organizations, the fundamental idea behind HyperCities is that all histories "take place" somewhere and sometime, and that they become more meaningful when they interact and intersect with other histories. Through the Google Maps and Earth APIs, HyperCities essentially allows users to go back in time to create, narrate, and explore the historical layers of city spaces and tell stories in an interactive, hypermedia environment. Partner teams have developed content connected to Los Angeles, New York, Berlin, Rome, Ollantaytambo, Tehran, Mexico City, Wellington, and many more places, big and small, urban and rural.

The project asks two seemingly simple—but deeply fraught and often contested—questions that are fundamental to identity: **Where are you from? What used to be here?**

The answers, of course, are far from simple or straightforward. As a globally-oriented platform that reaches deeply into archival collections and links together a wide range of media content (including photograph archives, 3D reconstructions, historical and user-created maps, oral histories and videos, GIS data, and community stories), HyperCities seeks to transform how humanities scholarship is produced, accessed, and shared, thereby yoking it to a public mission to document and delve into places in time.

When a user first visits Hypercities, what is shown is Google's satellite imagery of the world zoomed out to show the "historical cities" featured in HyperCities. There is no "starting" or "ending" point—just a digital globe and the chance to explore and contribute. Each time the user moves the map (zooms in, pans, jumps to a new city) or adjusts the time-bar, the application interacts with one or more external servers to retrieve relevant data based on the spatial and temporal bounding coordinates. I likened this process to archaeological coring, in which a sample might be taken from a single block or street over, say, 500 years, or perhaps a single city over a day. Such a search and query process was intended to simulate how visitors navigate a city with which they are not familiar: you proceed down a given street, you look around, and eventually you may even get lost. The flâneur in the age of new media becomes a kind of digital dandy, navigating a world that is open, under construction, and, to a certain extent, unknown.

Beyond browsing, HyperCities became a tool for leaving your own traces. With an emphasis on participatory learning, users can add media objects, curate and share their own collections, as well as view and link to other people's public collections. Users are able to add "micro-annotations" by geotagging locations in time, such as Unter den Linden in the year 1793, or the northwest corner of Leipzigerstrasse and Wilhelmstrasse from 1920 to 1945. The rationale is that these annotations contribute to the creation of a "people's history" of the city, leveraging the democratizing possibilities of the web to create, display, and distribute information. These annotations function collectively as "folksonomies," which complement academically generated taxonomies or "expert" content.

The birth of Web 2.0—the writerly web—has been well articulated by technology gurus such as Tim O'Reilly as well as leaders in the field of Digital Humanities such as Cathy Davidson and David Theo Goldberg, both of whom have advocated for a more open and participatory humanities, what might be called "Humanities 2.0."[10] Such a term refers to the generative humanities, a humanistic practice anchored in creation, curation, collaboration, experimentation, and the multi-purposing or multi-channeling of humanistic knowledge. It places a primacy on participatory scholarship, open-source models for sharing content and applications, iterative development, and interdisciplinary collaboration. In so doing, new communities—academic and the general public—are involved in the production of scholarship. This collaboration and interaction is at the heart of the HyperCities idea.

2009: AJAX and Web services

(TP) The HyperCities system architecture also follows one of the central trends often identified as Web 2.0: the front-end is almost entirely separated from its back-end. Although a web-based platform, HyperCities behaves more like a desktop application because the front-end follows an event-driven programming model rather than a standard webpage submission model. The platform is a collaboration of web services, compiling digital content from disparate sources through the use of XML, KML, JSON, and Javascript. The Google Maps/Earth APIs define a set of JavaScript objects and methods, which HyperCities uses to put maps on its interface, allowing instant integration of satellite imagery with other layers such as markers, pathways, images, historical maps, 3D objects, and other kinds of geo-data.

The technical goal of HyperCities is to be a generalizable, easily scalable data model for linking together and publishing geotemporal content using a unified front-end delivery system and a distributed back-end architecture. HyperCities consists

---

AJAX

```
$.post("./provider/objects/4023", metadata, _updateReturn, "json");
```

(DS) In the midst of a war for browser market share, Microsoft released a now-forgotten browser called Internet Explorer 5. One of IE5's most memorable features was a new JavaScript object: XmlHttpRequest. This weirdly-named object (it rarely has anything to do with XML now) allowed a JavaScript developer to request new data without reloading a page. And thus, AJAX (an even greater awkwardness of a backronym, "Asynchronous JavaScript and XML") was born, and with AJAX, Web 2.0: the entire fleet of Google apps, including Google Maps. And HyperCities. And social media, too: AJAX makes Facebook feasible (or would you rather reload the entire page every time you click "like"?), and finding your favorite restaurant on a map in Yelp (click to zoom out; reload the page).

And yet, AJAX is an awkward beast. From a meaningless name to an inconsistently-implemented feature set to a constantly-evolving set of standards, AJAX has incredible

of a geotemporal markup server and a front-end visualization platform built on the Google Maps/Earth APIs that enables users to explore, manipulate, and contribute to any geographically aware environment. At its core are databases of openly accessible, geotemporal content represented by KML, a markup language chosen because its development is funded by private enterprise (Google) but governed by the Open Geospatial Consortium, which ensures a robust user-base and an open-source development model for specification and implementation. HyperCities generates real-time, KML-based network

---

power yet odd responsibility. With AJAX, web pages grew into web applications. And now much anxiety is spent dealing with the potential for security holes: one web page may not make a request to a site on another domain—a page loaded on hackerparadise.com may not (and should not) make a request to gmail.com. This holds back developers who now see the web as a paradise of open data easy for computers to parse: bringing Twitter and Flickr streams together is far more difficult if your page cannot read from both of these domains.

Ajax makes the modern web possible. It allows a developer to display a web page, get more data, and update the page without reloading the entire page. It causes web developers to think about sites as bits of data to expose in a variety of formats—HTML, JSON, and XML. Or perhaps, it makes the web less the web and more the operating system: the web is no longer a set of pages of information (Berners-Lee and Nelson's vision) but a set of interfaces to data. APIs matter more when there's no page for a search engine to read. Thus, perhaps, AJAX has made the web more interactive and less human-readable.

links connected to geotemporal content, offering a non-exclusive front-end for contributing to, organizing, and exploring independent repositories. While HyperCities hosts and stores some data locally, one of the central aims of the project is to host metadata connections to content stored and maintained in external repositories and on external servers. These servers range from commercially available platforms (such as Google's 3D warehouse, YouTube, and Flickr) to library and archival platforms for maps, GIS data, oral histories, videos, photograph collections, and other media files. In this way, HyperCities provides connective tissue for the community of geospatial time travelers by leveraging the extensive development of data repositories and social networks.

The server back-end (written in PHP and running off a MySQL database) is limited to pulling new data to display and

---

From the Docuverse to the Dataverse: APIs

```
response = simplejson.load(urllib.urlopen(
            "http://search.twitter.com/search.json?q=obama"))
for tweet in response["results"]:
    for tag1,tag2 in itertools.combinations(r"#([\w\d]+)", tweet["text"])):
        tag_network[tag1][tag2] += 1
…
communities = louvain(tag_network)
simplejson.dump(communities, response)
```

(DS) The opening ceremonies of the 2012 Olympics contained a performance called "Frankie and June say 'Thanks Tim'," a tribute to Tim Berners-Lee's invention of the World Wide Web, which brought the eponymous protagonists together. At the end of the show, a house in the center of the stage lifted to reveal Berners-Lee himself at a computer. One wonders if, as he waved to the crowd, what passed through his head was that the messages Frankie and June sent on their smartphones were not part of the original web he had

inputting any changes a user might make to the objects being displayed. The front-end is almost a complete application itself because it contains all the display logic. We might think of it as a viewer for certain data formats, the same way KML can be viewed in Google Earth, ArcMap, and a host of other GIS applications. It is not only fairly easy to use HyperCities with different data sources, but it is also possible to pull the data from the back-end into any geographically aware environment.

2011: Social Media Mapping

(TP) Social media refer to a broad category of participatory communication and information technologies with (potentially) global diffusion of information sharing and data exchange. While it's easy to dismiss social media (Twitter, Facebook, YouTube, etc.) as superficial and amateurish, it is

- -

imagined: Berners-Lee's web was a set of documents people read with their eyes, not with their iPhone apps. What the web looks more and more like now is a series of APIs, in which data is exposed for machines to parse and interpret for humans.

APIs, or Application Programming Interfaces, are standards by which programs or web sites expose data for other programs' consumption. Facebook, Google Maps, Yelp, Twitter, and Flickr (to name just a few) expose their data through APIs. The smartphone apps market and social media could not exist without APIs; one of Twitter's biggest uses is applications built on its own API. With the proliferation of APIs, we can make the claim that data is much easier to repurpose than software.

But what can you really do with an API? All APIs have

becoming evident that they can be harnessed both singularly and in aggregate to fundamentally transform participation and even the notion of the public sphere. Part of the reason is that social media differ profoundly from traditional broadcast media such as newspapers, radio, television, and film by dint of their decentralization of authority, public accessibility, diffusion, ease of use, immediacy, scale, and, oftentimes, ephemerality. As the 2011 protests in North Africa and the Middle East have made clear, social media have changed the ways in which news and information are disseminated across the world and raised fundamental questions about who constitutes the public sphere, how we know and characterize historical change, and how we develop critical lenses for assessing social media that neither reduce them to epiphenomena nor naively celebrate them as the realization of digital democracy.

---

designs that consciously limit their purposes. We cannot harvest tweets more than seven days old, presumably so that Twitter's engineers can limit the amount of data the API ever has to access to a predictable amount, instead of the incomprehensible sublime of all tweets. Google Maps targets developers implementing "store locator" features: it assumes developers will want to identify a limited number of points, and does not implement GIS standards for exchanging large volumes of data. APIs also come with restrictions that govern the access of the company's data through a set of windows into the Master's house.

Many APIs suffer from a technical flaw, a feature I call the search-focus model. Most follow the REST (REpresentational State Transfer) model, proposed by Roy Fielding. In REST, each "thing" (a tweet, a Facebook status,

Since 2011, our team has developed a contemporary mapping project called "HyperCities Now," which geolocates, streams, and archives social media data from providers such as Twitter, YouTube, and Flickr. The project began with the protests in Tahrir Square in late January of 2011: what if we could listen to, map, and amplify the voices in the streets of Cairo in real-time and, then, archive those voices for scholars to study in the future? To do so, "HyperCities Now" began to stream and capture real-time social media feeds related to events from places around the world. We have archives of more than 450,000 tweets from Egypt over about 18 days of the Revolution, nearly a half-million from Libya, 660,000 in the weeks following the earthquake and tsunami in Japan, and hundreds of thousands from other events around the world, ranging from Anzac Day 2011 in New Zealand to the

---

a picture) is a "resource," and has its own URL, which allows it to be accessed and modified. Developers discover resources through a search and then focus on one resource to read or modify. This granularity has limits: it's harder to retrieve or access multiple items at once. It takes multiple queries to a RESTful API to alter the locations of multiple photographs, or to retrieve the profiles of many users who tweet with a particular hashtag. REST, a beautiful form in Fielding's dissertation, provides the developer with a small toolset. REST can be both an opportunity and a challenge to developers, restricting the scale and scope of their ambitions. Ironically, this increases the load on companies who serve APIs.

All this results, indirectly, from Berners-Lee's original work. He started us on the road to REST by

2011 riots in London, and most recently, millions of tweets from Hurricane Sandy, the Boston marathon bombings, and the 2013 protests in Turkey.[11]

Unlike broadcast news in which a reporter "goes on location" to retrieve the story, social media allow users to broadcast their own experiences in real-time from a place they are currently in. In this regard, it creates a new public sphere of voices and listeners who would never have been able to "hear" one another. We archived about 40,000 unique voices coming out of Egypt over the 18 days of protests and more than 200,000 unique voices from Japan. As an interactive, qualitative visualization, the goal of HyperCities Now is to foreground the ephemeral individual voice as it flashes up on the screen every few seconds, but it also gives users the power to listen to and analyze a staggering cacophony of voices

---

developing the protocol that these APIs use, HTTP (HyperText Transfer Protocol). In HTTP, a resource (like a webpage, or a tweet), has a "URL" (Uniform Resource Locator), which can be accessed using a "method." Berner-Lee's original specification, HTTP 0.9, had only one method, GET: readers' browsers would GET a page for them to read. Documents arrived on the web through protocols beyond the web itself.

HTTP 1.0 added the "POST" method, which made the web interactive. POST allowed users to POST data to an interactive program masquerading as a web page, and get back content based on what they had submitted—such as airplane tickets or a new blog POST. POST made the web interactive, and most content creation through APIs happens through POST. (Nowadays, we have twenty-five other methods, but GET and POST are the most

through geo- and time-encoded data. At the project's core are values central to the next wave of digital humanities: harnessing new technologies to expand the global public sphere, animating the archive in new ways, and using technologies to increase the impact, relevance, and importance of humanistic ideas in an ever-shifting world stage.

2013: Endgame—The HyperCities API

(TP) HyperCities has evolved into two components: a storage engine for geotemporal narratives and a browser for viewing these narratives. The API began as a separate project for accessing data within HyperCities using other clients. As such, the API offered the potential to repurpose the data to work with collaborating groups who wanted to import new data without using the geo-browser and/or read it from other

---

widely supported.) GET made users readers; POST made users writers and consumers, or whatever we might call those who submit (another strange word at the bottom of web forms) data for Facebook to track (the consumed consumer?).

Instead of web pages for human beings to read, RESTful APIs make web sites into interfaces to data. APIs matter more when there's no page for a search engine to read, or the data is locked down behind corporate walls. Now, we build applications to trawl through APIs for information. As the web has become more interactive, we require more technology to interact with the web. This is no longer a docuverse, but a dataverse.

clients. Over time, the idea of completely separating the back-end from the front-end became a goal, as it seemed to be a sensible way of preserving the data and allowing others to develop customized user experiences. In other words, we wanted users to be able to create, populate, and maintain their own instances of HyperCities. Ultimately, these multiple instances might form a "constellation" of sites that knew

---

The HyperCities Network/Nexus

http://hypercities.ats.ucla.edu/provider/collections?text=procession

http://hypercities.ats.ucla.edu/provider/collections/23077

http://hypercities.ats.ucla.edu/provider/collections/23077.kml

http://hypercities.ats.ucla.edu/provider/earth/collections/23077

(DS) If you use something like the URL above, you can search for data or download the data from a specific collection in HyperCities. You can even modify it, provided you have the privileges. You can download your collection as a KML file, edit it, and re-upload it. All this is due to the magic of the HyperCities API.

APIs make data consumable in other forms. A mobile application, another website, a desktop application, and the like can all read and edit the same data source. The public interface of HyperCities, what you see when you launch the application, is just one interface to our storage engine.

Consuming and providing APIs, HyperCities is part of a network. On the consumption end, we acquire map imagery from Google as well as ArcGIS and WMS-compliant map servers. On the producing end, HyperCities can export collections as standard KML files or as KML tours to play in Google Earth. HyperCities is a storage engine and an interface, and the

about and recognized one another, allowing data to be found and shared across the constellation.

While motivated by an intellectual goal, our repurposing of the API began in earnest around the time that Google announced that version two of the Maps API would no longer be supported. Google Maps version three has a completely different architecture from version two, and we realized

---

two are designed to communicate only through a specific interface, namely—the API. HyperCities is a nexus of parts, connective tissue between APIs. Why reuse code this way? Using others' APIs requires us to write and maintain less code, but it also creates a fragile network of dependencies, a house of cards in code. HyperCities is both part of a network of interrelated tools, and a network itself.

And yet, when APIs change, the cost of the network becomes apparent. Attempting to upgrade to version three of the Google Maps API has taken months and months of developer time; it was less an upgrade and more a rewrite. We didn't succeed, and so we decided to divorce HyperCities from its dependence on Google by allowing the developer community to create instances of HyperCities that run on any map front-end.

Even with all these points of failure, networking APIs make each part less vulnerable. With the API, even if we fail to do the upgrade, data will still be accessible through the server API. Google Earth should still read the tours HyperCities produces. Modular parts both make a network possible and also insulate themselves from too much exposure to the network; this networked, API-focused design negotiates a tension between independence, interdependency, and addiction. HyperCities is all of these things.

the upgrade would require a massive rewrite of the HyperCities application. Because the front-end relied on components we could not control and which changed all-too-frequently (the Google Maps API and the Google Earth plugin), the project was inherently fragile. We realized that another way to preserve the data against changes in the front-end was to make it possible to present the data in a variety of formats. This shifted the problem of preservation from one of maintaining the visual representations of HyperCities data to opening it to a variety of visual representations using different tools. It also allowed members of the developer community the opportunity to set up, administer, and customize their own HyperCities instances.

As the final stage of the HyperCities project, the team released the source code to the storage engine and the browser as open source projects, available on GitHub, a public source code hosting website, with a fully documented RESTful API. Among other things, it may function as a robust KML editor for geotemporal content. We imagine the final life of HyperCities will be as a publicly-available tool that can be installed on any server for accessing and storing geotemporal data and associated narratives. Users will have access to the API for developing their own clients and modifying the storage engine to work with other geo-browsers, something that will help to preserve data stored in HyperCities by making the data easier to port between formats that each browser can read. As such, developers may create new instantiations of HyperCities and thus make new kinds of geotemporal narratives.

With this consumer-agnostic API, HyperCities data is no longer a tenant of only Google's world; a developer may write a program for viewing a HyperCities collection using code and imagery from another provider. For example, the open-source mapping library OpenLayers can incorporate imagery from OpenStreetMap licensed under the Open Data Commons Open Database License. We have also written an OpenLayers-based browser, which has been released as an open source project and provides an opportunity for creating a HyperCities "liberated" from Google. Even if the master's tools cannot dismantle the master's house, separating these two components gives us our choice of masters, and allows us to open our doors to others.

Thick Mapping in the Digital Humanities

(TP) Until recently, mapping in the humanities was deeply bifurcated between what might be called, on the one hand, a "quantitative" approach using data analysis and visualization techniques adopted from the field of Geographic Information Systems (GIS, for short), and, on the other, what might be called "metaphorical mapping," variously articulated in cultural studies through theorizations of space and place, critiques of spatial systems, and critical cartography studies. The first is often dismissed as "positivistic," as uncritically importing methods of the social sciences into the interpretative and critical domain of the humanities with insufficient regard to the ideological biases of such information and visualization systems. At the same time, the second is dismissed by practitioners of spatial analysis on the grounds that it never actually engages with any spatial methods or mapping tools, neither designing environments for analysis nor creating "humanistic" systems for probing spatial relations.

Over the past couple of years, blended approaches have started to emerge in the digital humanities, which situate and investigate historical questions on spatial platforms, without uncritically embracing or cavalierly dismissing GIS. Richard

Marciano and David Theo Goldberg's "T-RACES" project, for example, brings together the history of redlining maps produced by the Home Owners' Loan Corporation in the 1930s with archival documents linked to census tract, in order to reveal the complex ways in which exclusionary spaces were created throughout the US to preserve racial homogeneity.[13] Here, a massive archive of American racial history has been geo- and temporally marked-up in a discovery and historical visualization platform built on the Google Maps engine. Another project, "Digital Harlem," also a historical mapping archive, was developed by four historians at the University of Sidney (Stephen Garton, Stephen Robertson, Graham White, and Shane White) "to visualize and explore the spatial dimensions of everyday life in Harlem during its heyday, 1915–1930."[14] To do so, the team created a series of dynamically populated, searchable map layers organized into three meta-categories: people, places, and events. Users can mix and match materials on historical maps and discover correlations, patterns, and trends that would not be discernible without tools for spatial analysis. Other projects—such as the US Holocaust Memorial Museum's Darfur genocide map and George Mason's Archive of Hurricanes Katrina and Rita—utilize the Google Maps/Earth platforms to document and investigate contemporary crises with participatory dimensions for the contribution of data.

Beyond Google, OpenLayers, an open source mapping platform with a global developer community, and Harvard's WorldMap, an ambitious, open source platform for exploring, visualizing, and publishing geographic information, are setting new standards for the preservation, dissemination, and analysis of networked geo-data. At the same time, major "spatial research centers" crossing the humanities, architecture, urban planning, geography, and design have begun to crop up, such as Harvard's Center for Geographic Analysis, Stanford's Spatial History Project (part of its Center for Spatial and Textual Analysis), Duke's Wired! Group (focused on dynamic visualizations of the past through modeling and design of urban environments), Columbia's Spatial Information Design Lab, MIT's Senseable City Lab, UCLA's Experiential Technologies Center, and multi-institutional collaborations such as the Virtual Center for Spatial Humanities. Each supports a range of spatial projects built on interdisciplinary methodologies coming from computational analysis and GIS, interactive design and four-dimensional modeling, information visualization, and statistical processing—all of which are part of the burgeoning field of "digital humanities."

The HyperCities project is part of this blended trend. HyperCities essentially allows users to examine the historical layers of city spaces and tell stories in an interactive, hypermedia environment. The central theme is geotemporal argumentation, an endeavor that cuts across a multitude of disciplines and relies on new forms of visual, cartographic, and time/space-based narrative strategies. To date, HyperCities features rich content on more than two dozen world cities, including more than 400 georeferenced historical maps, thousands of user-generated maps, and tens of thousands of curated collections, narratives, and media objects created by users in the academy and general public.

The Spatial Turn in the Humanities

(TP) When Edward Said published *Culture and Imperialism* in 1993, he justifiably maintained that "most cultural historians, and certainly all literary scholars, have failed to remark the *geographical* notation, the theoretical mapping and charting of territory that underlies Western fiction, historical writing, and philosophical discourses of the time."[15] According to Said, cultural criticism needs to "affirm both the primacy of geography and an ideology about the control of territory" (78). Why? In order to fully appreciate and critique the spatial matrix of modernity constituted by the broad forces of nationalism, imperialism, and colonialism. Modernity, for Said, is not merely a temporal indicator or periodizing concept but very much about specific colonial histories of power and control exerted on bodies in places.

In Said's wake, much work has been done in the humanities—particularly in transnational and post-colonial studies—to examine the "spatial strata" of cultural production and power: one need only think of studies such as Paul Gilroy's "black Atlantic," Arjun Appadurai's "global ethnoscapes," Homi Bhahba's "location of culture," James Clifford's anthropological study of "routes," Saskia Sassen's and David Harvey's studies of the global diffusion of capital, Fredric Jameson's "geopolitical aesthetic" and (somewhat under-theorized) notion of "cognitive mapping," Franco Moretti's map-based analyses of the "spaces of literature," Stephen Greenblatt's call for "mobility studies" to focus on questions of diaspora, exile, and displacement in literary and languages studies, not to mention renewed attention to psycho-geographies, imaginary landscapes, and practices of *détournement* rooted in situationist ideas of urban engagement and cognitive dissonance. And in the wake of Derrida and Foucault, poststructuralist notions of the instability of signifying practices and critiques of power dynamics have been rigorously taken up in the expansive fields of critical cartography studies and geography, as the work of J. B. Harley, Edward Soja, John Pickles, Denis Cosgrove, and others attests. And more recently, as both of these fields have deepened their interaction with visual studies, digital humanities, historical GIS, and information visualization, new areas of research have been catalyzed at the nexus of mapping technologies, politics, and design through the work of scholars such Laura Kurgan, Trevor Paglen, Anne Knowles, Annette Kim, Bill Rankin, and many others.

While this "spatial turn" in the humanities has clearly fostered many differentiated approaches to analyzing and mapping the imbrication of culture/power/space, what has been missing—at least up until recently—is a fundamental rethinking of *the medium* in which cultural criticism and historical investigations are undertaken. In other words, we cannot adequately study spatial systems, mapping conventions, landscapes of power and control, colonial networks, histories of emplacement and displacement, cultural flows and hemorrhages, sites of memory and oblivion without considering the media in which to map the complex interplay between lived and experienced spaces, representational spaces, and imaginary spaces. In fact, if we cast this problem more broadly, what's at stake is the attention to media, materiality, and method in humanities scholarship, the very fundamental ways in which cultural-historical questions are articulated, investigated, and emplotted as arguments.

Bringing together spatial-temporal narratives, visual design and argumentation, embodied navigation, and curatorial strategies to imagine new modalities of engaging with the past and the present, these issues have been catalyzed by the digital humanities subfield variously called "Spatial Humanities," "digital cultural mapping," or simply "thick mapping." What if the many, competing pasts saturated in a single place could be mapped onto and along streets, neighborhoods, and territories? What if, following Edward Casey, culture was reconnected to place and the movement of bodies in space and through time?[16] To do so, we need a methodology and a medium that foregrounds time-layers, or sedimented palimpsests, such that histories become proliferated as intertwining layers, making it possible to tell more than one story at the same time, or any number of possible stories. This does not mean that "anything goes" or that "what actually happened" no longer matters; instead, it gives way to richly interactive, multiplied stories in which the singularity of narrative succession has been abandoned in favor of extensible digital spaces that variously map, intersect with, and also disrupt physical, embodied spaces.

Zoom and Thickness as Historical Methods

(TP) Let's look at a Pharus map of Berlin from 1926. It is an extraordinarily detailed, five-colored, ink-printed map of Berlin, in which prominence is given to significant German architectural monuments and transportation networks (including railways, subways, and express roads). The former are represented as miniature, three-dimensional models, rendered in black ink and dusted with a brownish-gold tint: the Reichstag, the Victory Column, the Royal Castle, the Berlin Cathedral, and the Kaiser Wilhelm Memorial Church, among others. And over the map, we see railway stations rendered in red with their connecting lines spread out across the city. Interestingly, the main stations—Potsdamer, Anhalter, and Lehrter—have accompanying signs indicating their possible destinations both in Germany and beyond: Cologne, Frankfurt, Munich, and Hamburg, but also Paris, Basel, Vienna, and Saint Petersburg. Not only is Berlin connected to an international network of cities throughout Europe, but one can travel to these cities in any order one wants: from the Anhalter train station, Basel comes before Leipzig and Munich comes before Dresden. As part of an interconnected network, they do not demand a definitive order or a unitary direction; instead, they can be experienced in any number of new temporal and spatial configurations. (Figs. 4 / 5)

Walter Benjamin surely saw maps like this, maybe even this one, as Pharus printed the most popular folding maps during the 1920s and 1930s, and is especially well-known for its pocket-sized maps of Berlin for the 1936 Olympics. In fact, Benjamin references a Pharus map in passing in his city montage, *One-Way Street*, a text that was completed in the same year, 1926, and uses the street as its organizing principle.[17] But unlike the Pharus Map, in which one can actually follow the procession of a street much like the movement of a flâneur, Benjamin's urban meditations do not easily map onto the city or its traditional, cartographic representations. In *One-Way Street*, for example, the narrative begins quite zoomed in at the street level, at a filling station, before moving into a breakfast room, the Number 113, a visit to Goethe's house, memories of Chinese curios, the Mexican embassy, and a construction site. While many of these vignettes may have been derived or distilled from the vast signifiers of the urban landscape—"this space for rent," "Optician," "Lost-and-Found Office," "Post no bills!" and so forth—these signifiers certainly do not add up to produce a map that looks anything like the Pharus map. Instead, they offer various optics for seeing, remembering, and narrating the urban space, optics that are a function of experience, perception, and zoom level. For Benjamin, zoom becomes a historical method for narrating the city: zoomed in means close reading and close analysis of details, whereas zoomed out means macro-level, comparative perspectives. Zoomed out, one discerns structures and patterns of the city as a whole but compromises texture and precision; zoomed in, one sees locally but loses the view of the whole. Of course, there is no reason to privilege a single level of zoom; rather, we ought to consider "zoomability" as a way of investigating space, one that is further enabled by (but not limited to) digital mapping and, thereby, considers history as places to be mapped, as sites for the emplotment of narratives as cartographies.

At the same time that he made use of zoom, Benjamin was also concerned with things unrepresented on and absent from the Pharus map: social and economic structures, childhood memories, emotions, fleeting images, tastes and smells, noises, textures, and other somatic experiences. One map is not "more realistic" or "more accurate" than the other; instead, we have different interpretative and symbolic systems for representing the spatio-temporal order of modernity: a planimetric map and a mental map.[18] Through various strategies of selection, visualization, and interpretation, both produce the space that they ostensibly represent. The Pharus map is part of a spatio-temporal order that stretches back at least to the eighteenth century and has become naturalized in its cultural redundancy, utilitarian value, and political effi-

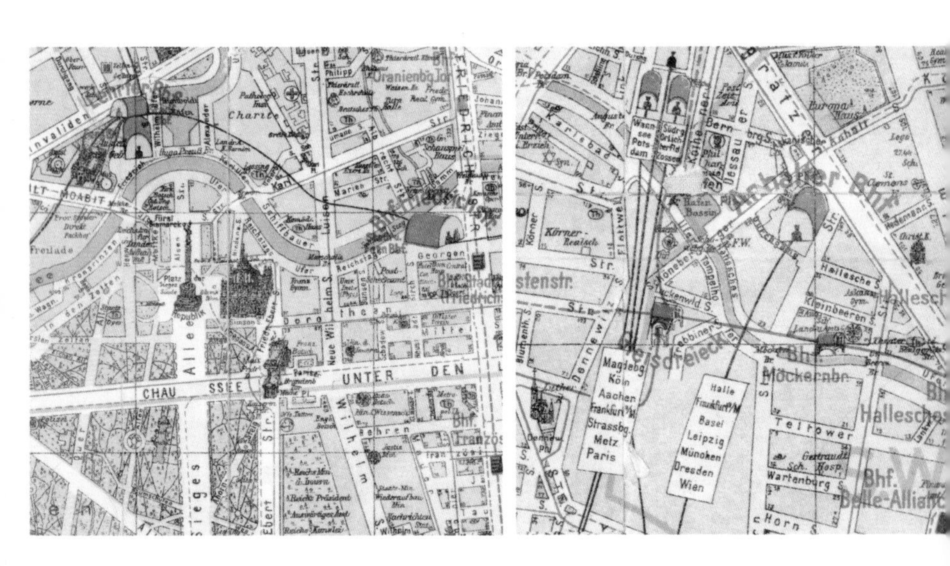

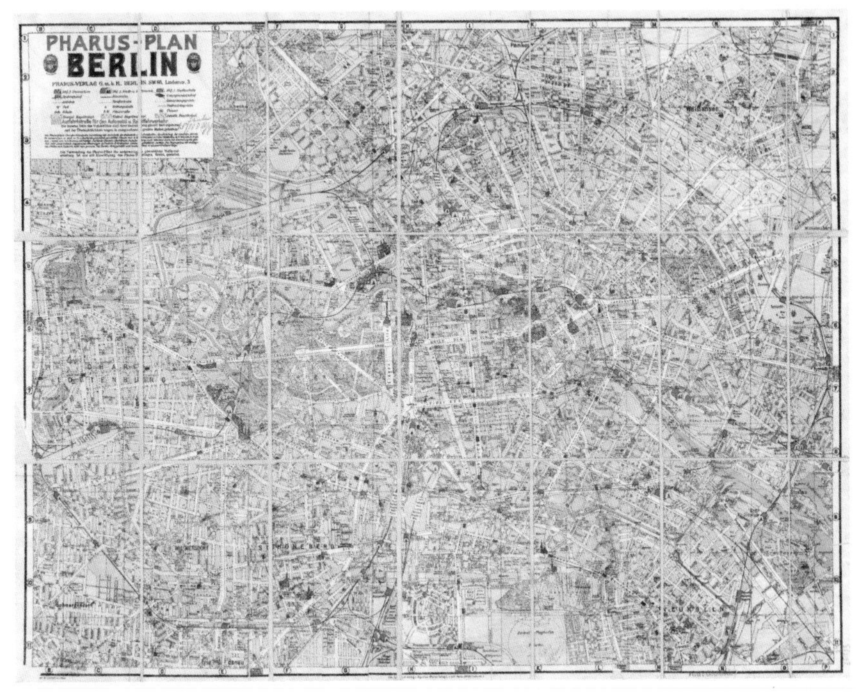
PHARUS-PLAN
BERLIN

(4) / (5)

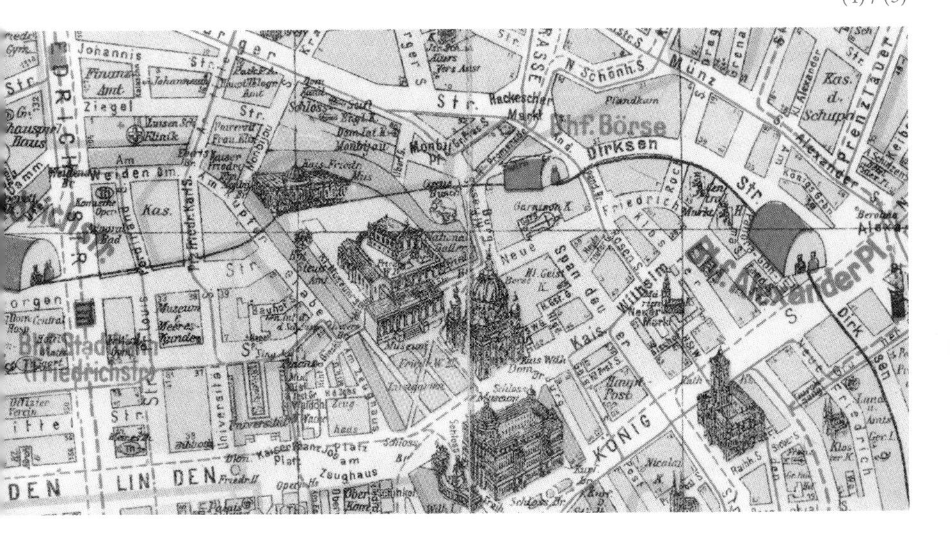
Johannis
Ziegel
Kas.
Hackescher Markt
Bhf. Börse
Dirksen
Friedrich
Neue
Kais. Wilhelm
Bhf. Alexander Pl.
Universität
Zeughaus
Museum
KÖNIG
LIN DEN

cacy: we have become used to looking at maps with a bird's eye perspective, a grid-like organization of streets, framed boundaries, a clear coordinate system, planimetric accuracy, and alignment to true north.[19] The birth of cartographic reason is inextricable from the history of the territoriality of the nation-state, both its internal linkages and expansive—imperial and global—ambitions.[20] The critical question remains: how can such a map (and, thereby, the construction of its history and the history of what it represents) be deconstructed and re-animated—that is to say, opened to the infinite number of non-simultaneous histories contained in every street, structure, and building, the innumerable voices and bodies that made these histories through their interactions and contingent encounters in such spaces?

While the Pharus map abstracts and excludes the kinds of experiences that Benjamin privileges in all of his city reflections, it is organized by the logic of a particularly modern spatio-temporal system in which the temporality of experience and expectation are both bound up and broken apart from one another. Every significant architectural monument and building is depicted in exactly the same way, as if to cast them all as part of the permanence of the past or the inheritance of history, what Benjamin calls, unsympathetically, "their enshrinement as heritage" (*Arcades Project*, 473). The past is given value because it is inherited, and every structure, even the most recent, is endowed with the temporality of the oldest (in this case, the Royal Castle), resulting in a kind of leveling effect in which the non-simultaneous becomes simultaneous. What is far more relevant for Benjamin's cityscapes and a central part of the "thick mapping" imagined here are the non-simultaneous, fractured histories that co-exist as "time layers" in any given present.

And there is another, decidedly modern spatio-temporal logic operating on the Pharus map: namely, the reconfiguration of space and time ushered in by the railway system. It would have made little sense to indicate Paris, Basel, Vienna, or even Leipzig and Hamburg on a Berlin map prior to the mid-nineteenth century. While these places could obviously be approached by carriage (something that easily took days), one didn't think of them as places already "in" Berlin. On this map, the railway stations are like worm-holes: one goes in at Berlin and pops out in Paris. It is not coincidental, then, that Pharus privileged the railway system on the map, as there was arguably no technology of modernity that had more of an impact on the reconfiguration of space and time than the railway. In 1843, with the opening of a number of major rail lines around Paris, the poet Heinrich Heine famously declared railways to be "providential events" because they "killed" space and intimated the coming death of time.[21]

The "new time" (*Neuzeit*) of modernity was both a break from the eschatological temporality of the past and the institution of a new, world standard time, something precipitated as early as 1842 in London with the creation of railway timetables. Acceleration, progress, and speed became the mottoes of the modern world, leading to the construction of an interconnected, globalized world in which Paris could be in Berlin and vice-versa. The material superstructures of modernity made of iron and glass were built to showcase transcendental size, speed, and mobility, but they also harbored this world's destructive capacities. The railway system is a fundamentally dialectical construction, embodying the hopes and dreams of the nineteenth century as well as the horror and catastrophes of the twentieth.

This dialectic is something that punctuates Benjamin's writings in exile, as he constructs a commemorative, imaginative geography of German places no longer occupied by

Jewish bodies. In 1932, having already left Germany, Benjamin imagines setting his life out on a map. Exile becomes a site of displacement from which to imagine another history as well as to map out his own life. In the early drafts that he made for his chronicle of Berlin, he writes: "I have long, indeed for years, played with the idea of setting out the sphere of life—bios—graphically on a map. First I envisaged an ordinary map, but now I would incline to a general staff's map of a city center, if such a thing existed."[22] He then goes on to mention things he would mark in his "system of signs" in a kind of classification scheme or legend: houses of friends and girlfriends, assembly halls, hotel and brothel rooms, benches

in the Tiergarten, prestigious cafes, and what he calls "street images" from "lived Berlin" (597). The idea of setting these out in a General Staff's map is striking, as these kinds of maps were produced for military campaigns to illuminate the topographical features of the landscape, including information about populations and transportation networks. As far as I know, no such map of Benjamin's life has surfaced, although one could certainly argue that all of his experimental writings on travel and urban space (from the early city portraits to the *Arcades Project*) were attempts to not only map his life but also to think through what it might mean to write history in graphic form, to map culture and spatialize history, to bring together the experience, representation, and production of space.[23] It is an avowedly non-mimetic, anti-developmental, non-linear model of imagining history as places to be mapped, one which is rooted in exile, displacement, and the disembodiment of the German/Jewish experience.

(6)

In recollecting "images" of his childhood in Berlin, Benjamin pauses on one train station in Berlin in particular: the Anhalter Bahnhof. When it was rebuilt and reopened in 1881, it was the largest, most expensive, and most opulent station in Europe. He writes: "The Anhalter terminus [refers to] the mother cavern of railroad stations, as its name suggested—where locomotives were at home and trains had to stop [*anhalten*]. No distance was more distant than when its rails con-

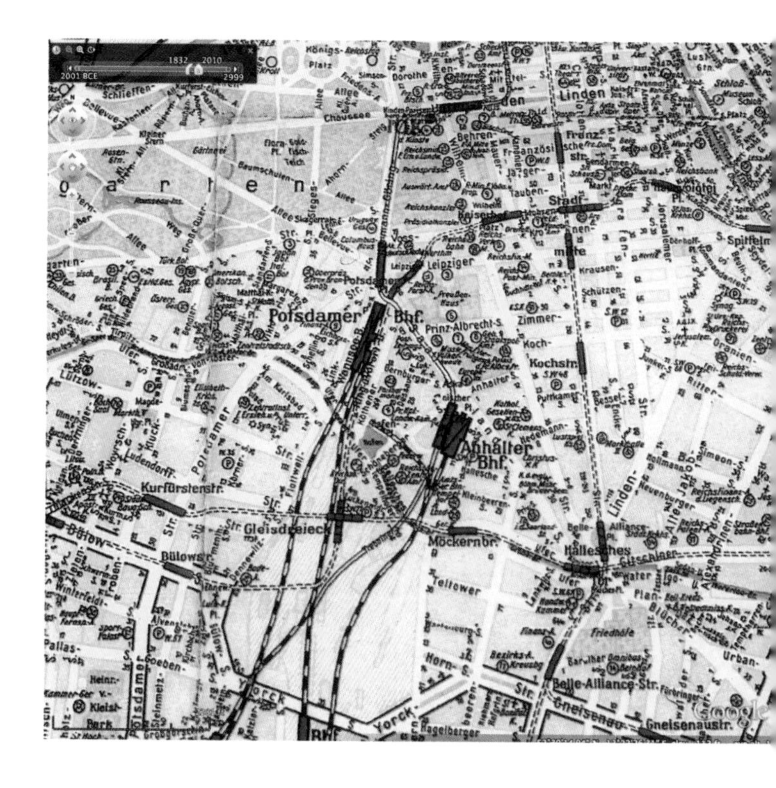

verged in the mist."[24] To Benjamin the railway was the reality of that marvelous and equally dubious nineteenth-century dream of progress characterized by, among other things, the possibility of connecting to a faraway place; it is where he recollects leave-taking from the city and arrival back at his childhood home. The Anhalter was also where Franz Kafka arrived from Prague when he visited Felice Bauer in Berlin; it's where Paul Celan stopped over on his way to Paris from Czernowitz on the day after Kristallnacht. In the 1930s thousands of Jewish children were sent on trains from Berlin's Anhalter Bahnhof to safety outside of Germany; and in 1941–42 the station was used to gather elderly Jewish "transports" who were de-

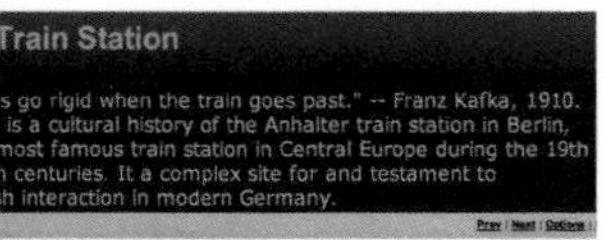

(7)

ported to the concentration camp of Theresienstadt. Although the iron and glass roof of the station collapsed during one of the last bombing raids of Berlin, the station was not completely destroyed, and, after the war, trains began running again as of August of 1945. They continued to run until 1952 when the tracks were cut by the division of Berlin and later by the erection of the Wall. After much debate, the ruined station was razed in 1961. Most of its remains were disposed of in the early 1960s, except for part of the front portal and the southbound railway tracks. These tracks were left to the forces of nature since their last use on May 17, 1952. For more than five decades, birch trees grew between the ruined tracks. It was not until 2008 that the urban wasteland between the former Anhalter and Potsdamer train stations was finally cleaned up and reclaimed by the city of Berlin for another future. (Figs. 6 / 7)

In this highly constricted but thickly layered place, one can move diachronically—much like archaeological coring—through a remarkable band of German/Jewish history, tracing contingent moments of encounter, interaction, mobility, and destruction. Although no longer visible, each time-layer coexists in this stratified place, from which one can move forward or backward. At the same time, one can also proceed synchronically, stopping at a particular time and moving horizontally through space, noting the closeness of the Anhalter train station to other layered structures and streets in Berlin, such as the Gestapo Headquarters on Prinz Albrechtstrasse in 1944 or Felice Bauer's home in 1912. In this sense, the time-layers of the Anhalter station open downward as well as laterally, calling up the possibility of an infinite number of stories and encounters—in other words, thick maps. Like so many fraught, overdetermined places in Berlin, one discovers, in Benjamin's words, "the crystal of the total event" (*Arcades Project*, 461), the dialectic of modernity.

What would it mean, then, to produce a cartographic history of modernity, not simply a history of modernity in maps but rather a practice of history that was spatial, a way of understanding events and cultural encounters by plotting them thickly onto maps?[25] This line of thinking opens up an investigation of how modernity is not just a temporal designation (as in *Neuzeit*) but also a practice of cartographic reasoning, spatial representation, and geographic persuasion and control. We might call it *Neuraum* ("new space"). It was Michelet, after all, who famously declared that "history is first of all geography."[26] And so by "thick mapping," I mean this quite literally: creating and interrogating maps, time-layers (*Zeitschichten*), spatial imaginaries, and geographies of movement and encounter in order to create geotemporal narratives that follow the expansive and particular spatial logic of a railway map, not unlike the "worm-hole" on the map that connects Berlin to Paris at the Anhalter Bahnhof in 1926.

If one takes space (rather than time) as the prerequisite of historical narrative, it becomes impossible to write unidirectional, developmental stories; instead, there is a nearly infinite proliferation of perspectives, stories, interactions, and possibilities. What would it mean to produce narratives that looked more like railway systems or webs, with a multiplicity of connecting segments, branches, nodes, and possible pathways to get from "here" to "there"? The result is a labyrinthine structure in which straying and contingency are the methodological starting points. The necessity of chronology, progress, teleology—or just the gentle, forward movement of a historical argument—gives way to spaces of possibility. It makes little sense to speak of "before" or "after" or necessity as a modality of movement; instead, we get temporally layered, spaces of possibility, marked by distance and proximity, contingency, simultaneity, and networks of connection. Not only can readers or viewers insert themselves at multiple points and look for their own orientation but the narratives themselves are multilayered, fractured, and open-ended. Thick mapping begins to look like an ever-expanding railway system that moves out and downward at the same time, giving rise to a participatory web of intersecting cities, voices, streets, memories, and narratives. This is the "humanities idea" behind the HyperCities digital mapping project.

W
1

Los Angeles Ghost Maps

Ghost Map:
Geveronga-Yaanga / El Pueblo de Nuestra Señora Reina de Los Angeles / La Placita / Downtown / Los Angeles, California, circa 0 C.E. to 2000 C.E. Cartography and data by Phil Ethington, 2010.

/ La Placita / Downtown / Los Angeles, California, USA, 0 CE - 2000 CE

Every act recounted by historians literally took place. All human actions and experiences take place, in the sense that, embodied, they cannot be placeless, but also in the sense of seizing the site of their occurrence and marking it for generations to come. It becomes a *topos*, a reference-point in the vast network of human knowledge. Making places, people inscribe their material landscape with meaning. To the extent that these inscriptions endure, places are

Map Call-Out 1:
Figueroa Specters, 1935, 1940, 1997.
Top row: Figueroa between Temple and Sunset, underneath the Four-Level Interchange, looking west. Bottom Rows: same location, looking east. Middle Row: Houseless community dwelling beneath the Four-Level, 1997. 1935 photography by the Los Angeles Bureau of Engineering. Watercolor and ink drawings by the Works Progress Administration (WPA) and the Los Angeles Planning Department, 1940. 1997 Photography and photomontage, Phil Ethington, 1997.

dynamic accretions of meaningful actions. If the past is the landscape of what took place, then writing history, as representing, recounting, and narrating the paths of our ancestors in that landscape, is inherently cartographic.[27]

These observations lead to a re-conception of historical interpretation as the act of reading places. History has always seemed so obviously to operate in the field of time and temporality, that its essential spatiality has long been obscured, especially in the era of modernity, which valorized the processes of time over the habitations of space. Since the spatial turn, we are now able to see how processes also and inescapably inhabit and inscribe. Once we recognize the spatiality of the past, its form becomes a matter of inescapable interest. Pictorial representations of social forms can be thought of as the mapping of the footprints of past actions. Past actors inscribed and left the scene. Their inscribed actions haunt every subsequent inhabitant as ghosts, shaping lives invisibly. The historian who maps the past makes these ghosts visible.

Each place is a unique configuration of power and people who share living ecologies with the natural world. In *Ghost Metropolis*, I narrate and explain the *topoi* through which we can envision the *pasts* of Los Angeles, California, U.S.A. The topography of a region is the topology of the emplaced, place-making people, carving their own nodes of power and freedom within the grand institutional landscape that they inhabit. Because many prior generations erected these coral-like structures of institutions (of labor, of discourse, of cultural, residential, racial, ethnic, gender, labor, and wealth segregation), all new stories to tell are built upon older stories. I tell these stories with a combination of verbal textual narratives and cartographic depictions. Cartography as a pictorial form operates by *simultaneity* and *juxtaposition*; verbal text is syntactically linear and narratological. As such, each form of communication can do something that the other cannot. Together, we can say far more than we can with either alone.

Ghost maps are hand-crafted

Map Call-Out 2:
Ruins of Boylston St, 1940s-1990s.
This montage depicts the north side of Boylston at the top of the street seen in Ghost Neighborhoods, with the detritus of former lives found by the author and his collaborator in the basement foundations. Aerial photography, Fairchild Airphoto, 1945, courtesy of Whittier College. Watercolor and ink drawings by the Works Progress Administration (WPA) and the Los Angeles Planning Department, 1940, Los Angeles City Archives. Photography and montage, Phil Ethington, with Steve Appleton, 1997.

composites of archival analog cartography and vector-based, digital GIS layers. They are designed visually to reveal time, change, events, and motion through the symbolic languages of color, shape, iconography, and textual annotation. They are dense, thick maps, not made to be read quickly, as would be mere illustrations or diagrams. As a rich and complex graphical composition drawn directly from the profound complexity of past social life itself, the content of a ghost map—like the metropolis itself—exceeds the capacity of any textual narrative to explain it. Each ghost map, crafted in pictorial language, is a free-standing document, offered for the reader to ponder and to puzzle over, to return to many times.

As the critical history of cartography has made clear, maps are as imaginary as they are objective. When composed by the historian to recount the past, however, the geometric and the metaphoric can be deliberately coordinated to achieve that which the art of historical narrative has always sought: to move and persuade the reader/viewer. Alongside my textual narratives and ghost maps, I compose photomontages, panoramas, and mosaics of photographic stills and other forms of graphic depiction. These compositions are both *denotative*—showing factually the where and the when, and *connotative*—implying meaningful relations among the visual components. As "call-outs" or "wormholes," these photographic compositions are also indexed to points and shapes in the ghost maps. Visualizations of complex historical phenomena, as they took place, these compositions draw much of their representational power from the indexical nature of photography—an empirical record of the past.

The Ghost Map portrays the central and most layered districts of the global metropolis of Los Angeles—a very old work-in-progress, retaining the shapes of power inscribed by many generations since the first human settlements of the late Pleistocene Era. Conquered by the Uto-Aztecans circa 0 CE, the settlement on the banks of the Los Angeles River where it enters the Los Angeles Basin, was conquered again by the Spanish Empire in 1781 and re-established as El Pueblo de Nuestra Señora Reina de Los Angeles (The Town of Our Lady Queen of the Angels). Centered on La Placita (the Little Plaza), this settlement became the nucleus of the mighty Los Angeles metropolis. The toponyms used here are those inscribed by each successive regime—each society who ruled this place. The "base map" (bottom-most layer) is a 1925 US Geological Survey (USGS), which (using early aerial photography as source material) recorded the locations of every built structure. Contour lines visualize the landforms.

All social interactions *take place*, leaving institutional deposits. But all societies and individuals are constantly in motion. The census data series on this Ghost Map tracks the changing racial-ethnic and social class composition of the neighborhoods from 1940–2000. As a work of both interpretive humanities and geographic social science, the Ghost Map is accompanied by an elaborate set of legends, indexing known data to the precise polygons from which those data were collected. Readers can follow a statistical narrative through stacked bars (racial composition, denoted by colors arbitrarily assigned) and pie charts (ratio of blue to white collar workers), each indexed to the census tracts over seven decades.

The freeways, constructed in this Downtown area from 1948–1953, obliterated many residential spaces and created new transportation flows and new barriers to street-level circulation. The "Four-Level Interchange" [See Call-Out 1: "Figueroa Specters"] created a new meta-intersection next to the arterial intersection of Sunset Blvd. and Figueroa Blvd. Edward Doheny's "City Field" [See Call-Outs 2 and 3: "Ruins of Boylston St. and "Ghost Neighborhood"] brought petroleum to the surface just west of downtown, later overbuilt with residential neighborhoods, which in turn were razed for urban renewal, and then for a high school.

The houseless inhabitants recorded in Figueroa Specters created their community beneath the mighty Four-Level Interchange, the first high-speed concrete freeway in the world, finished in 1953. Constructed by a Cold War developmental regime that trampled democratic rights and razed the homes and neighborhoods of thousands of Angelenos, this massive structure inscribes new forms of injustice into the metropolis. Joseph Guajardo (far right, middle row, Figueroa Specters), one of these houseless men, was born and raised in one of the homes within sight of his open-air bed where he slept circa 1997, when his photograph was taken. The Ghost Map traces these paths of destruction with the transparent, ghostlike freeways, the homes of Guajardo and his neighbors, visible underneath. Ghost maps preserve the visibility of older, prior shapes beneath those that have taken their place in later times.

Phil Ethington

Map Call-Out 3: Ghost Neighborhood, Beaudry and Court Streets, 1940, 1997. These open fields just west of Downtown were once vibrant neighborhoods, razed in the 1980s to make way for a development that never materialized. In the late 1990s, the remaining foundations and other remnants of former homes were graded away to make way for a Los Angeles Unified School District mega-project, an ill-planned, $300 million high school, millions of which were spent to seal the school from poisonous gases that a competent school board would have known about before attempting construction there. Sited atop the former oil field of petroleum magnate Edward Doheny, methane gas rises to the surface, haunting students while they train for a better future. Watercolor and ink drawings by the Works Progress Administration (WPA) and the Los Angeles Planning Department, 1940. Photography and montage, Phil Ethington, 1997.

PDub Productions: Mapping HiFi

Clutching his electric guitar, eyes closed, Angelo "Manok" Bernardo mimics the Jimi Hendrix poster behind him. The soundtrack is Hendrix's version of the "Star Spangled Banner," only it's Manok, a recent Filipino immigrant, playing the song. "Manok" stands for chicken, he explains. As a young boy he had a fighting rooster and started drawing chickens. The name stuck. These scenes open "Anthem," Manok's video from *PDub Productions*, an innovative media and civic engagement program in which local youth create digital content about Historic Filipinotown for distribution on multiple platforms. "Anthem" cuts between Manok's story and his efforts to bring Filipino Christmas caroling into Los Angeles' Historic Filipinotown, aka HiFi. As "Anthem" closes, Manok plays a Hendrix-style version of the Philippines' national anthem, "Lupang Hinirang."

"Anthem" is Manok's personal story; yet it's the neighborhood's story, too. Both are full of contradictions, hybridities, and conundrums. Designated as such in 2002, Historic Filipinotown has few cultural or physical markers to indicate the area's status. It is better known for its proximity to the

http://www.hypercities.com/pdub

101 Freeway and as a shortcut to downtown than as a cultural nexus. Although only 2.1 square miles, the significance of HiFi extends far beyond its borders. Today, HiFi is ethnically and culturally diverse, with Filipinos comprising 20% of the population and Latinos the majority.

Historic Filipinotown is different from other Los Angeles ethnic "towns." HiFi just looks like Anywhere, Los Angeles. The richness of the neighborhood is not so much in the visible structures, but more in the people, culture, and the history. The larger Filipino American community is much like HiFi in that we are not very visible. We are spread out all over L.A. Due to our great ability to adapt, we tend to blend in. Because we are not visible, many are surprised to learn that Filipinos are the largest Asian population in the city of Los Angeles.

Aquilina Soriano-Versoza, Executive Director, Pilipino Workers' Center

Enter PDub Productions, aka PDub.

PDub was designed to deepen a sense of place—not simply to describe, document, and depict that place. By engaging generations of community members in the process of representing HiFi through its own stories, it aimed to help shape HiFi's future. It introduced young immigrants, new to the neighborhood but also to the country, to HiFi. It taught them the digital media skills they didn't learn in school. It deepened their interest and investment in their new community by directly introducing them to people who work in, on, and influence the community, while giving *PDub*'s youth authority as HiFi's media makers.

As such, *PDub* addressed several community needs: it built a sustainable digital media program for resource-poor neighborhood youth; it engaged local youth in creating content about Historic Filipinotown's character, history, and significance; it convened an inclusive, multi-generational, cross-sector coalition of neighborhood stakeholders; and it enhanced the neighborhood's visibility and foot-traffic. *PDub* uses digital technology in a creative, educational, and participatory manner to advance ongoing efforts to benefit the community-at-large. Its programs appear on multiple platforms and in many forms: as interactive Mobile HiFi Tours, on mapping platforms such as Hyper-

Cities, as street-level multimedia installations, and even on metro buses. The goal is to create a new type of cultural landmark—one that is participatory, iterative, reflective of the community, and malleable in format.

Partners

PDub was designed to blur institutional boundaries and disciplines, pairing small organizations with giant ones. The Pilipino Workers' Center (**PWC**) was the community anchor and hub. *PDub* grew out of PWC's desire to create a program for Filipino immigrant youth and give them access and entry both to the community and to media-based skills. **Public Matters**, an interdisciplinary, social enterprise consisting of artists, media professionals, and educators, developed the partnerships, drove *PDub*'s focus, designed the curriculum, led after-school workshops, and did post-production of all media-based work. *PDub* was HyperCities' most in-depth exploration of a community in Los Angeles. In the HyperCities instantiation of the project, content was created by community members—particularly youth—and exists side-by-side with content created by scholars, cartographers, and the government census bureau. The project was generously supported by the MacArthur Foundation/HASTAC, through one of its first "digital media and learning" prizes in 2008. Finally, UCLA's Center for

Research in Engineering, Media + Performance (REMAP) further expanded *PDub*'s media reach to the general public by assisting students in designing the framework for a time-travelling Mobile HiFi Tour, writing the tour's code, and helping to create *PDub*'s largest piece of mobile media.

> PDub *shaped me as a person and as an artist. Projects were predominantly youth-led, from conception to completion. PDub projects, such as re-imagining and recreating a* harana *in English, composing original music, developing and shooting a music video—daunting tasks all of them—were not tasks I could expect in school.* PDub *demanded confidence, vision, and collaborative effort. I gained confidence through the trust I was given in my ideas and guidance by teachers (more like mentors) who nurtured my wanting to challenge mainstream aesthetic sensibilities.* PDub *put faith in our collective ability to create thoughtful art via filmmaking.* PDub *was my introduction to that art form, and it instilled an interest I am currently developing and nurturing into a career. I stand endlessly grateful.*
>
> Xochilt Sanchez, *PDub* Student

PDub and power

PDub's partnerships flipped many pre-existing power structures, whether institutional, educational, or age-related. *PDub* youth were primarily recent Filipino immigrants and local Latino high school students. Of the students who typically go to college in the neighborhood, most attend vocational school or community college. Many enter the work force as caregivers or enlist in the military. In *PDub*, however, they drove the intellectual content and scope of the project. They were behind and in front of the camera. They became the community experts and educators. They worked directly with artists, community members, and UCLA students and faculty through a seminar called *"Creating and Recreating Historic Filipinotown,"* taught by UCLA history professor Jan Reiff. As a collaboration between UCLA students and *PDub* high school students, the *PDub* youth were thrust into the position of teachers, becoming the local informants and experiential archive. The team's community mapping projects are "published" on the HyperCities platform and provide complex, intersecting insights into the layered and all-too-often erased histories of the neighborhood.

http://hypercities.com/LA

Taking mobile media in new directions
In order to reach a broader audience, we pushed the notion of mobile media both in concept and in scale. We did performances. We held indoor and outdoor screenings. We created GPS-enabled walking tours and large-scale transit tours. As a result, *PDub*'s audience grew to include community activists, L.A. history buffs, elected representatives, business owners, and immigrants riding the bus.

> *PDub Productions, PWC's Jeepney, and the Mobile HiFi Tours all have the goal of raising the visibility of the Filipino American Community. The tours have created tangible pathways to experience the Filipino community and HiFi. The jeepney represents Filipinos: it's bright, it's colorful, and it's unmistakable. With the Mobile HiFi Tours you can literally drive through and delve into the richness that's here.*
>
> Aquilina Soriano-Versoza

The *Mobile HiFi Immigrants' Guide* brought people to HiFi. Rather than a conventional guide focused on key dates or sites, the stories and perspectives of generations of newcomers to Los Angeles shaped the user experience of place. Like HyperCities, the *Guides* were portable, time-travelling devices that took people through space

and time. The *Guides* were divided into four time periods that started in the 1880s (when Filipinos first came to Los Angeles) and ended in the present. Each period featured a central figure, told in a first-person narrative created by community members, and a clickable period-specific map.

PDub's mobile media literally grew larger and larger. Jeep Willys were among many things the U.S. Army left in the Philippines after WWII. Filipinos transformed them into flamboyant forms of public transportation. PWC acquired an original 1944 Sarao jeepney, tricked it out with a GPS unit, a flat screen monitor with digital maps, a microphone and sound system, and a special version of the Mobile HiFi Guide. Cruising through HiFi in the iconic orange PWC Jeepney as crowds cheer is an unforgettable experience in creating and learning about community memories. And thanks to L.A. Freewaves' "Out The Window" program, *Pdub* went mobile on the Metro buses, too, reaching an audience of millions weekly. Many bus riders are working class immigrants. *PDub* featured student stories of immigration and a series entitled "Hidden HiFi."

PDub and digital mapping

PDub was a dynamic interactive program and process. But what does it have to do with mapping? Everything. We created physical mappings, inspired by the power of place, and digital mappings that

can be endlessly combined and recombined to explore stories of displacement and emplacement, location and relocation. They all bear witness to social, political, and economic geographies of power through the dynamics of push and pull that shape neighborhoods and communities. In HyperCities, the video histories created by *PDub* youth exist side-by-side with five decades of census and demographic data (such as median income levels, home values, ethnicity, and education levels), rendered as GIS maps in order to let quantitative ("social science") data speak to, interact with, and be enhanced by qualitative ("humanistic") stories. They form new discursive registers through their interactions and gain historical depth and complexity when further situated on historical maps, such as the 1939 "redlining" map of Los Angeles. The "redlining" of what would later become HiFi includes a description file from the Home Owners' Loan Corporation that claims that the neighborhood is shifting to "subversive racial elements." It is precisely through "thick mapping" that such frictions become part of the consciousness of the present and a motivation for on-going education and civic engagement.

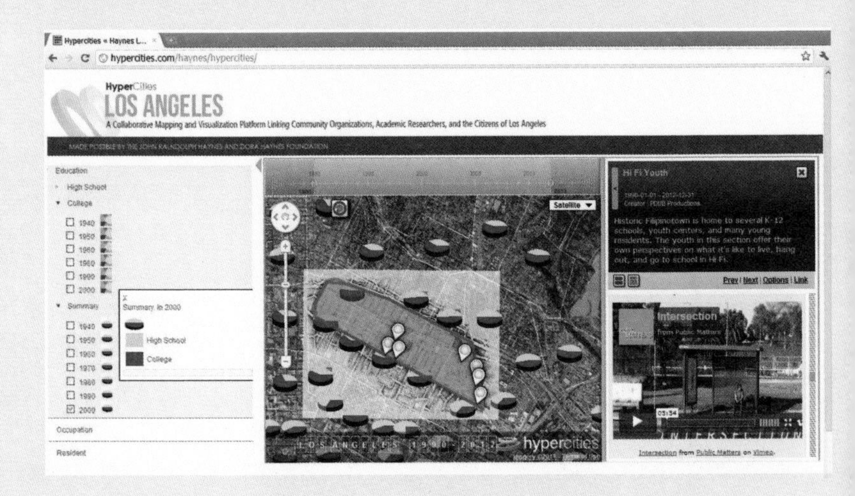

The process of raising the visibility of Filipino culture and the Filipino American community also helps to reconnect and deepen the understanding of Filipinos themselves. Filipino immigrants often do not know about the important struggles and victories of the Filipino farmworkers. American-born Filipinos were never taught in school about the Golden Age of Filipino American Boxing and have felt a sense of always being an "other" whose face and culture are rarely reflected in the mainstream telling of America. The visibility through PDub *has helped the Pilipino Workers' Center to start construction on a new building with 45 units of affordable housing. The stories of the Filipino workers helped to build a sense of dignity and pride in Filipino caregivers who formerly were ashamed of their positions. It gave them confidence to educate legislators about the need for equal rights and protections for domestic workers. And it helped educate a generation of youth with the skills and knowledge for understanding, mapping, and finally transforming the city they live in.*

Aquilina Soriano-Versoza

Mike Blockstein and Reanne Estrada,
Public Matters

The View from Above / Below

Toward a Media Archaeology of Google Earth

(TP) Viewing the world in Google Earth has naturalized a set of habits (flying, panning, zooming) linked to a long-standing desire to rise above the earth and look down from the perspective of the Heavens. This desire can be traced back to Greek and Roman mythology when the figure of Apollo's detached, omniscient eye first took hold as a cipher for flying above and seeing the earth.[28] The fantasy of external spectatorship of the earth is deeply wed to the history of empire, the rise of the nation-state, and the colonial will to know, domesticate, and control space. It can be traced from the mythologies of antiquity and the cosmographies of early Christianity to Renaissance and early modern cartographic visions of the whole earth and, later, rationalized notions of a fully enlightened earth taken to their logical conclusion with the modern desire to accurately map and know every point on the planet. Transcending time and space, Cicero's dream of Scipio ends with his liberation from the planet as he surveys the universe and gazes back at a small planet earth. Centuries later, Immanuel Kant, conceives of the sublime—*das Erhabene*—as unbounded elevation, of a mental state of soaring above the ground, liberated from the clunky facticity of the human body. Today, military satellites encircling the earth survey movement on the ground down to centimeters, while social media applications running on GPS-enabled mobile devices transmit and share the geographic and temporal coordinates of a user's every move.

But it was not until 1968 that the earth was actually seen (not imagined or pictured) as a whole by human eyes with Apollo 8's escape from the earth's orbit and its subsequent lunar encounter. This was the first time that human beings saw the earth from the depths of space. The famous "Earthrise" photograph taken in December of 1968 documents this spectatorial encounter, showing the earth partially enveloped in darkness next to the lunar landscape. Although often re-oriented and widely circulated to depict the lunar landscape along the bottom, the original Earthrise photograph is quite unsettling since it radically disrupts the attempt to re-ground human perception.[29] (Fig. 9) If you zoom all the way out of Google Earth, you see a diminutive image of the earth—perhaps not coincidentally—from almost the same perspective and distance as this photograph, but while sitting firmly and safely ensconced in front of your computer screen. (Fig. 10) In 1972, during the last lunar mission, another photograph of the earth was taken, this one simply called NASA Photo AS17-22727, which depicted a geometrically framed planet earth inscribed in a black square. (Fig. 8). The photograph follows the Tropic of Capricorn and foregrounds Africa and the Antarctica (Cosgrove, 261), while other such photographs variously "re-center" the earth with the North Atlantic on top. Whereas the Earthrise photograph seems to betray a somewhat modest point of view by capturing the receding earth in its smallness, contingency, and distance, the 1972 photograph seems to declare itself as triumphal: a perfectly geometrical, completely rationalized image of control and even imperial hubris. This is the realization of Heidegger's fear of the "the world conceived and grasped as a picture": the earth is merely an image of an astronomical mass, conquered, rationalized, and presented by the human subject.[30] As a whole earth viewer, Google Earth oscillates between the world-views and ideologies of these two images.

Google Earth valorizes external spectatorship on the totality of the earth, the pleasures of flying, and the instantaneity of seamless, hassle-free travel without leaving the safety of a home computer. But it is also a media form which has to be understood as a spatial-cultural practice that has a long history in the West connected to knowing, organizing, seeing, and controlling the world, on the one hand, and unknowing, disorganizing, obscuring, and subverting the world, on the other. While it is tempting (and even quite plausible) to see

(9)

Google Earth as the culmination of the totalizing, transcendental gaze, it is also a wildly participatory platform for decentering this gaze, for playful uprootings and particularist story-telling, something that we have spent a lot of time exploring with the HyperCities project. In other words, Google Earth is Apollonian and Dionysian at the same time.

To begin to unpack this cultural form, we might consider Google Earth from the standpoint of media archaeology, namely the media-specific, discursive practices and topoi, especially visual ones, which form the conditions of possibility for representing and seeing the earth like this.[31] How did seeing and flying in Google Earth become a culturally redundant and even naturalized habit of viewing and navigation? When did it make sense epistemologically? In other words, what had to happen for me to believe this perceptual experience as a natural, correct, and true experience of viewing and navigating the earth? What does it mean, after all, that "the earth" is imagined and manipulated as a whole? Here, we are not just interested in Google Earth as a technology but want to emphasize how it is a spatial-cultural practice with a significant pre-history related to a multiplicity of media and cartographic forms for viewing and representing the totality of the planet earth. Crystallized by the Apollonian desire to see as a transcendental spectator, Google Earth is the product of many different perceptual habits and histories of visuality, including:

- The desire to travel to faraway places without leaving home, something that is emblematically embodied in the history of the nineteenth century panorama, and lives on today with the treatment of reality like a spectacle for the safe and privileged viewing of the West, something that, in Susan Sontag's critical words, "universalizes the viewing habits of a small, educated population living in the rich part of the world."[32]
- The military history of Global Positioning Satellites, remote sensing and remote seeing, the history of aerial reconnaissance, and the bombardment of cities throughout the twentieth and twenty-first centuries.
- The history of media as forms of navigation, stretching from the first reconnaissance photographs taken from balloons and airplanes to city films in the late 1920s, flight simulators and contemporary videogame interfaces structured as navigation through virtual spaces.
- The cartographic fantasy of the completely surveyed and mapped earth, which found its early forms in sixteenth century world atlases that sought to condense knowledge of the physical surface of the earth into a flat representation, and which gradually closed the distance between the representation and the referent (a fantasy that Borges takes to its logical conclusion in his famous story of a map the same size and quality of the land it represents).[33]

(10)

- The expansion of capital into previously unknown, uninvented, or untapped realms ("the digital"), while simultaneously naturalizing itself and eliding its differential histories, inequities, and injustices.
- The Situationist aesthetic of *détournement*, of taking and cutting up maps, of overlaying them as montages in unsystematic ways and creating new forms of navigation, story-telling, and memory-making.
- An ethic of curation and responsibility to otherness, to people and places that have not, historically, been "on the map," and which gives rise to infinitely extensible digital spaces of participation.

As media theorist Lev Manovich argues, the cultural form of navigation is at the heart of almost every iteration of "new media" and can be traced throughout the twentieth century: it stretches from the first reconnaissance photographs taken from airplanes during wartime in the early 1900s to Vertov and Ruttmann's city films in the late 1920s, to the development of immersive flight simulators during the Cold War, to the creation of computer-generated animations and simulations in the 1970s, to contemporary videogame interfaces structured as navigation through virtual spaces as well as the embodied experiences of virtual navigation created by new media artists such as Jeffrey Shaw and Bill Viola.[34] Among many other places, we see this media form of navigation exemplified in Microsoft's wildly successful Combat Flight Simulator series, the various city instantiations of Grand Theft Auto, or the infinitely expansive virtual worlds of Second Life. The history of new media is wed to the history of the navigable space interface, the ways in which a spectator visualizes and experiences movement through space.

The navigable space interfaces created in the late-1990s and early-2000s by the ART+COM group for projects such as "The Invisible Shape of Things Past" can be regarded as one of the epistemological and media-specific antecedents to Google Earth and HyperCities.[35] ART+COM explored the fundamental concepts of panning, zooming, and overlaying (now so culturally redundant as to seem naturalized) through panoramic camera movements, still images, layered historical maps, and time-based media. Focused on Berlin after the fall of the Wall, ART+COM aimed to create a virtual city as a kind of archive of perceptual habits and media forms, which could be accessed over the web and eventually integrated with the physical landscape. The idea was not to replicate or extend the hyper-realism of contemporary computer graphics but rather to interrogate the media forms themselves through interfaces that probed both the temporal and spatial layers of the city and the perceptual habits of film and digital media.

But to understand the media-technology roots of moving through space without leaving home—a kind of virtual tourism—we need to turn back to the history of the panorama, a much longer technological and conceptual cultural form for panoptic viewing. The word "panorama," coined at the end of the eighteenth century, means "all-viewing" and was variously instantiated throughout the nineteenth century in large-scale, survey landscape paintings and photographs, installed on round or circular surfaces in buildings specifically designed for panoptic vision. The basic aim of the panorama, as Stephen Oettermann argues in his comprehensive history of the mass medium, was "to reproduce the real world so skillfully that spectators could believe what they were seeing was genuine."[36] Unlike traditional canvasses, these paintings completely surround and envelop viewers in an environment that functions as a portal or gateway to a faraway reality, be it scenes from historical battles, overseas places, or the totality of the metropolis. Housed in major European metropolises like London, Paris, Berlin, and Vienna, the panorama was a decidedly western perceptual mode of viewing the local as well as the foreign, in which the power of sight was deeply wed to the imperial conquest of real and imagined spaces. Most panoramas functioned by suturing the view of spectators to what was to be seen in such a way that the depiction of a panoramic reality passed in front of their eyes, such as with the numerous Kaiser Panoramas throughout Germany, which were functional for nearly a century. (Fig. 11)

However, the Great Globe in London's Leicester Square between 1851 and 1862 turned that mode of vision inside out: it allowed viewers to enter into the globe itself and see the continents and oceans painted in relief on the entire interior surface of the sphere, at a scale of 10 miles to one inch. (Fig. 12) Accessed through four viewing platforms, this immersive experience of viewing the whole earth from the inside out inverted the position of the transcendental spectator zooming

(11)

into the earth from above and, instead, placed this spectator at the center of a viewing experience from within (analogous to the "galaxy view" in Google Earth, when you look outward rather than down). In both cases, viewers are placed within an expansive scopic regime in which they are constituted as subjects of knowledge by virtue of a panoptic gaze.

In 2005, Google and Microsoft launched their own digital globes, Google Earth and Microsoft Virtual Earth, respectively. Using comprehensive imagery obtained over years by a host of US satellites, viewers running these applications can zoom into virtually any location on earth and look down, with often extraordinary levels of resolution, on the surface structures of the earth. With a broadband internet connection, anyone can transmogrify their spectatorship into the Apollonian eye, a synoptic and transcendental view that is unencumbered by the facticities of physical embodiment: one's locality—seemingly—ceases to matter; political, economic, or social circumstances—seemingly—no longer condition spectatorship once the connection is established; and

(12)

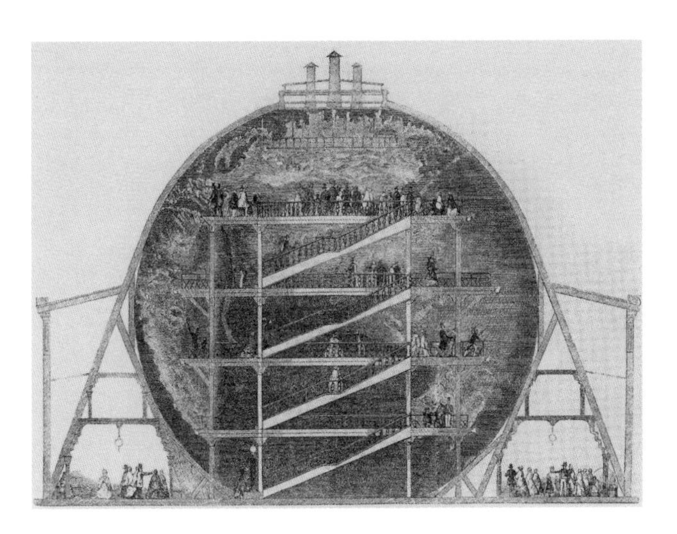

one's body is never put at risk. As Jonathan Crary has argued, this detachment of spectatorship from the body is actually the hallmark of a host of visualization technologies, which "[relocate] vision to a plane severed from a human observer ... Most of the historically important functions of the human eye are being supplanted by practices in which visual images no longer have any reference to the position of the observer in a 'real,' optically perceived world."[37] Severed from the human body, seeing and traveling in Google Earth is a safe and pleasurable experience in which movement is frictionless, borders are non-existent, and mobility is just about limitless. It is no coincidence that this experience of seeing brings about reverie, a kind of sublime feeling in which one is elevated above the earth in a dream-like state.[38]

Although the application is certainly much more than the imagery itself, it is all-too-often forgotten where the satellite photographs came from and what purposes Global Positioning Systems (GPS) and Geographic Information Systems (GIS) originally served. As Caren Kaplan has argued in her cultural history of "aerial targeting," the ways in which GPS and GIS produce militarized consumer and citizen subjects through discourses of precision often ignore, if not obscure, the military infrastructure that enabled these kind of subjectivities in the first place.[39] Kaplan and other critics such as Lisa Parks are rightly concerned about the intimacy between the "whole world ethos" of the 1960s and 70s, its imbrication with the history of the Cold War, and the televisual capacities enabled by GPS satellites, which "disembody vision and construct seemingly omniscient and objective structures of seeing and knowing the world . . . [that posit] the world (or the cosmos) as the rightful domain of Western vision, knowledge, and control."[40] The viewing technologies and satellite imagery of Google Earth have to be recognized as the product of warfare, particularly the Cold War and the modes of seeing made possible by remote sensing and global surveillance, aerial war, precision guided missile systems, and combat flight simulators. Indeed, Google's satellite imagery comes primarily from DigitalGlobe and GeoEye, two companies that own high-resolution, earth-observation satellites, and sell their imagery to customers, including the U.S. government, the military, and private industry, for purposes of intelligence, monitoring, defense, and situational awareness.[41]

It is hardly coincidental that early versions of Google Earth came with a pre-installed flight simulator for navigating its logistical space. Google Earth is simply the next generation of hyper-realistic flight simulators, a basic form of interactive media that uses three-dimensional graphics to navigate through virtual spaces that are closely modeled after the physical world. In his cultural history of Microsoft's *Combat Flight Simulator 2*, Patrick Crogan argues that military flight simulation has informed—structurally and technically—the development of computer-aided design, animation, and multimedia computer games: "The flight simulator game genre … is the direct descendent of computerized flight simulation developments that have been so crucial in the history of computer-generated imaging and simulation."[42] In its prevalence in mainstream culture through film, television, video games, and web applications like Google Earth, the flight simulator, as Manovich points out, has naturalized a way of seeing in which the street-walking flâneur is transformed into a combat pilot, often obscuring or forgetting "the military origins of the navigable space form."[43]

While Google Earth seems to offer a totalizing, immediately accessible view of the completely illuminated earth, this totality and immediacy is quite misleading, as the imagery is not a single snapshot of the earth but rather a complex, ever changing montage, created and re-created over years, of regions of the earth bathed in sunlight.[44] In the early history of cartography, there are significant forerunners to this kind of representation of the whole earth through compilations of geographic knowledge, most emblematically the amalgamated atlases produced and printed by Gerardus Mercator and Abraham Ortelius in the sixteenth century. Mercator's world map of 1569, *Nova et Aucta Orbis Terrae Descriptio ad Usum Navigantium Emendate Accommodata*, for example, was a montage of many information sources, including explorer reports, travel narratives, eyewitness accounts, and imaginative speculations, ranging from those of Strabo and Pliny to Marco Polo and Magellan. The map itself was produced in 18 sheets from copper engravings and included many of the sources as documentation on the map itself, with "unknown" areas in North America, the polar regions, and the southern hemisphere variously fictionalized and sometimes occupied by sea monsters, cannibals, and strange beasts.[45] Ortelius's 1570 world atlas, *Theatrum Orbis Terrarum*, arguably the first modern atlas, was also an amalgamation (or perhaps, a summation) of a diversity of information sources, cartographic representations (including his original 1564 atlas, *Typus Orbis Terrarum*), and sources that depicted the known and unknown world. Predating Google Earth by four and half centuries, both maps present the earth as a gridded whole, the product of European colonial encounters in which knowledge and sovereign power have become ever more deeply wed to produce a depiction of the earth as a whole to be surveyed, known, and controlled.

In the realm of digital mapping of the whole earth, the forerunner of Google Earth is the Van Sant map, the most widely reproduced digital image of the whole earth ostensi-

---

Where is (0,0)? Deep in the Ocean of Geo-data

(DS)

```
coords = this._getTagValue(xml, namespacePrefix +
"coordinates");
...
path = coords.split(" ");
len = path.length;
// If marker has only one coordinate pair, it's a path
if ( len === 1 || path[1] === "") {
    // Make sure we have Point Tag
    if ( xml.getElementsByTagName(namespacePrefix +
        "Point").length > 0 ) {

      bits = path[0].split(",");
      this._points[0] = new
          GLatLng(parseFloat(bits[1]), parseFloat(bits[0]);
      this._type = this.MARKER;
    }
} else if ( len > 1 ) {
    // Build the list of points
```

bly depicting an unmediated, totalized view of the planet from outer space. Distributed in the 1990s by the National Geographic Society as a poster called "A Clear Day," the montage is an engineering composite consisting of thirty-five million scanned pixels of the surface of the earth taken between 1986 and 1989 by National Oceanic and Atmospheric Administration satellites. One can see a strikingly similar image upon zooming all the way out of Google Maps: it is a pixilated world without clouds, smog, night, weather patterns, and people.[46] Just like the Van Sant map, the earth never existed like this! Yet, the image presents itself as a true and clear reflection of the earth as a whole, one that masks all friction, social structures, and even its own conditions of possibility.[47]

```
    for (i=0; i<len; i++) {
      bits = path[i].split(",");

         point = new GLatLng(parseFloat(bits[1]),
parseFloat(bits[0]));

      this._points.push(point);
      this._bounds.extend(point);
    }

    // It's a Polyline
    if ( xml.getElementsByTagName(namespacePrefix +
"LineString").length > 0 ) {
      this._type = this.POLYLINE;
    }
         // It's a Polygon
    if ( xml.getElementsByTagName(namespacePrefix +
"Polygon").length > 0 ) {
      this._type = this.POLYGON;
    }
  }
```

Unlike the overt constructedness of early modern atlases or modernist city novels and films (such as Alfred Döblin's *Berlin Alexanderplatz*, in which pieces of narrative are literally gathered from the trash of the city and stitched together as if contiguous, or Walter Ruttmann's city film, *Berlin: Symphony of a Great City*, in which the shock of the urban experience is betrayed by the cutting apart and splicing together of imagistic spaces of the metropolis), Google Earth seems to offer a mimetic, highly rationalized, and virtually frictionless mapping of digital space onto physical reality. One wonders if Horkheimer and Adorno were right: does the fully enlightened/mapped earth radiate disaster triumphant?[48]

(DS) Where is nowhere? Where does an object go when mapping code can't understand it? I'll tell you where: Off the coast of Africa, where the equator and the Greenwich Meridian meet, in the Gulf of Guinea.

Why does this happen? To understand, we need terms from that great computer scientist Stephen Colbert. Colbertian concepts of "truthiness" and "falsiness" are used to do mathematics on incompatible values. JavaScript—the language of web mapping—has a variety of data types, and allows a programmer to auto-convert between them, a process called "casting." In places where a numerical variable must be true or false, JavaScript assumes that a zero means false, and a positive value means "true." Truthiness is positive; falsiness is not. This is usually safe: an operation that returns a value of zero usually means "don't do anything."

How many ways can something be falsy? There's "undefined," "null," "false," and "zero." For numbers, there's also "NaN" ("not a number"). For a symbol to be undefined, it must never have existed; otherwise, the least-existent

But rather than insisting on and unmasking its constructedness, Google Earth has facilitated the march toward a fantastic realism, adding "reality" features that allow users to overlay present-day weather patterns, real-time lighting (including the rising and setting of the sun), shadows, three-dimensional buildings, terrain features, ocean views, 360-degree street views in panoramic bubbles (Fig. 13), and even the ability to see the stars and galaxies from the perspective of the earth on a clear night.[49] The physical limitation and usefulness of Borges's map growing so large as to finally cover the territory represented is no longer the issue: with Google Earth, physicality has been overcome by an infinite virtuality, as ever greater

- -

it gets is "null." After that, false and zero mean a value has been defined; they are assumed to be equivalent, and null is close enough. False == zero == null.

So what happens if Google Maps fails to understand numbers that represent a set of coordinates? All of its attempts to parse them result in NaN, which is falsy, or more falsy than truthy. Make NaN a number, and the closest number to NaN is 0. Therefore, your badly-formed numbers (misread your XML?) become 0s to Google Maps. Your points end up in the middle of the Atlantic Ocean, in the deep waters of geo-data, rippling in the digital waves off the coast of Africa, or stranded on "Null Island," as the makers of Natural Earth, a company that distributes public domain map datasets, calls it. Ultimately, it's the consequence of an epistemological order that was codified globally in 1884 when Greenwich was selected as the prime meridian, the point of origin for every longitude measurement and clock in the world. And so it remains today: the origin (0,0) is the site of the unrepresentable from which all other points are nevertheless measured.

viewing capacities, functionalities, and datasets can be loaded into the application. Where does it end? With a ubiquitous web of information (the semantic web) merging, inextricably, with the planet earth. This is the possibility—and perhaps also the horror—of the geospatial web. It is the essence of what might be considered Web 3.0, a Matrix-like fantasy of the all-pervasive web[50] that promises not to be evil.[51] As Paul Virilio presciently pointed out in his analysis of the logistics of perception in the twentieth century, the ultimate goal of the war machine (of which Google Earth is a comparatively benign product) is to become "an all-seeing Divinity, . . . a general system of illumination that will allow everything to be seen and known, at every moment and in every place."[52] This can only happen when the all-knowing information machine merges—in its transparency, ubiquity, and instantaneity—with the all-seeing war machine. One wonders how far away we are, or if we have already arrived.

And what, then, might we actually see lurking behind these images of absolute totality? Far from a utopian picture of global unity and harmony, the image of the whole earth reflects a body in pieces, a radically broken, fragmented, and inequitable reality marked by, among other things, radical disparities of wealth, social injustices, political strife, environmental disasters, and technologies of violence and destruction. The very representation of the earth from a distance—whether through Google Earth, the lunar mission photographs, or sixteenth century world atlases—gives rise to the dream of a unified totality, a global body that obscures the fundamental nexus of knowledge, representation, and power. For this reason, it is all the more urgent to deconstruct these images of global totality, to interrogate their conditions of possibility, to undo their silences and erasures, and to decolonize their knowledge systems. There is no unified, homogenous earth; there are only

(13)

spatial practices of representation that function as historically contingent ways of knowing, controlling, and possibly liberating.

(DS) To delve into these conditions of possibility, it is worth looking at the technical and corporate history of Google Earth a bit more. The original goal of both Google Maps and Google Earth was to give users local insight into the world around them, to transform digital navigation into physical navigation of nearby streets, rather than to provide a global perspective on the world. According to the design teams of both products, the local perspective was more marketable, and the global perspective was almost an afterthought. This tension between local and global levels persists in both tools.

Google Earth began life at Intrinsic Graphics, in a division spun off into Keyhole, Inc. as EarthViewer3D. Co-founder Avi Bar-Zeev cited a desire to give users a whole-earth perspective for educational and environmentalist reasons, motivated in part by Al Gore's Digital Earth initiative, which he considered as an inspiration for EarthViewer3D.[53] That said, EarthViewer3D only provided high-resolution imagery for the relatively short list of major cities on Keyhole's website.[54] The program offered low-detail imagery for the majority of the world at a scale of one kilometer per pixel (at this scale, California, Nevada, Arizona, and Utah all fit in the same window). The company's website promoted the tool to real estate companies, city governments, and a few federal and military customers, but given the level of imagery, it seems unlikely that EarthViewer3D would have allowed a user to see or accomplish much.

Projections

(DS)

```
var MapApplication = function () {
  var _externalProjection = new OpenLayers.Projection("EPSG:4326"),
    _internalProjection = new OpenLayers.Projection("EPSG:900913"),
    _gsat = new OpenLayers.Layer.Google("Google Satellite", {
      type: google.maps.MapTypeId.SATELLITE
    }),
    _instance = new OpenLayers.Map('map', {layers: [_gsat]});
    _instance.setCenter(new OpenLayers.LonLat(-0.11, 51.50)
    .transform(_externalProjection, _internalProjection));
}();
```

The importance of the global view emerged later in the development of Google Earth. The program's direct links to surveillance and national security are questionable, but do drift together in the public imagination. CEO John Hanke confirmed that his intention was to link the company's name to the Keyhole spy satellites, part of the CIA's Corona Program to photograph the Soviet Union and the People's Republic of China between 1959 and 1972.[55] While the company's website advertises homeland security applications, this is only one of many markets it mentions. In fact, Hanke attributed the company's growing number of federal customers in the early 2000s to the sobering post-9/11

- -

These eleven lines of JavaScript (or similar ones) preface any display of Google Maps imagery in OpenLayers. They are an illusion of an illusion. They create the comfortable bubble in which the Google Maps API automatically envelops all of its projects, and it is the bubble HyperCities lives in. Unless you step outside of the bubble into the wild world of GIS standards, a Google Maps developer never has to confront the ugly truth of cartographic projections, including the one right under the thick layers of JavaScript they use every day. A Google Maps developer drops data onto a spherical globe, a perfect 360 degrees. The satellite imagery that they are using, however, is a square, measured in tens of thousands of meters. Google's API converts degrees to meters so transparently that developers never notice that their sphere is a square; an OpenLayers developer must face the square and convert it to a sphere using the code above.

economy rather than Keyhole's own specific marketing efforts. EarthViewer3D achieved public recognition when news programs used it to display imagery of bombings during the Second Gulf War, but the company specifically added high quality imagery from Baghdad and Tikrit rather than allowing news agencies to use imagery already available in EarthViewer3D. As a commercial product, the global perspective was generally unimportant, and its ability to give the user a global perspective for defense purposes was limited.

Google Earth represented a significant change from EarthViewer3D: not only did Google Earth provide high-quality imagery for the entire world, it was free. Keyhole initially charged $599 for the program and required an annual subscription. This made Google Earth

---

(DS) To paraphrase William Gibson, a projection is a consensual hallucination, or a hundred: we agree on a fixed reference point and agree to measure all distances from that point. While setting a standard can seem cold—after all, there was no reason for Greenwich, England, to be the absolute longitudinal reference for dividing up the world—even cartographers and engineers can't agree, and so projections proliferate. The common World Geodetic System 84 (WGS:84) standard is a lie: the world is an ellipsoid, not a sphere. Based on WGS:84, the Spherical Mercator projection (EPSG:4326, EPSG:90013, or SR-ORG:6) is an act of violence, chopping off the poles at 85 degrees in either direction to make the world a perfect square. All are too big; a city cannot use a projection where the difference between one street and another may

a widely accessible tool for exploration and a platform for experimentation, as sites such as Google Earth Hacks and Google Sightseeing demonstrate.[56] It was not businesses and federal clients who wanted to see the entire world, but consumers who came to Google Earth out of general curiosity (ironically enough, often to see their home from the air). To both the engineers and businesspeople behind EarthViewer3D, a global perspective seems to have been a neat trick rather than a serious tool.

Globe demos are, in fact, a common test for image-rendering hardware, but these are not commercial products. Indeed, one such demo inspired Google Earth: Mark Aubin, a Keyhole employee, attributed his vision of the product to "From Space to Your Face," a simulation

---

be less than a thousandth of a degree. And thus, the cacophony of coordinate reference systems emerges as a chaotic field of possibilities. Take your pick: would you like Equal-area? Equidistant? And from which international body will you take your name? The Eurocentric, technocratic European Petroleum Standards Group? The private-sector Environmental Research Systems Institute? And would you prefer longitude, latitude or latitude, longitude? Pick one and stick to it. Your data lives in that world; but whatever world it lives in is just one of many. Since reference points and units cannot be fixed, no universal system can represent the world without compromise. All attempts to transcend that world result in creoles of coordinates and endless translation. Just as there is no pure language, there is no true projection.

that he had created at Silicon Graphics International several years earlier.[57] "From Space to Your Face" demonstrated the advanced graphics capabilities of SGI's Onyx workstation: it began with a global view of the earth, and then zoomed down to the SGI logo on a Nintendo 64. As the name suggests, it was graphics engineers rendering the entire world out of more Dionysian impulses than Apollonian ambitions.[58]

Google made the choice to develop the global perspective in Earth to extend an existing commercial product into a platform for data exchange. Global spectatorship, in turn, necessitates public openness: businesses and governments care less about a global view than a few important local perspectives, such as San Francisco, London, and Baghdad. Digital tourists are more acutely aware of gaps in coverage, and insist, in turn, on global coverage. Consumers' own demand for a seamless global vision has paradoxically led to exposing those seams.

Counter-Mapping

(TP) While Google Earth may have become the platform for realizing the long-standing desire to be a transcendental spectator looking down on the world, these same technologies have also helped to dislodge this spectator and repurpose the militaristic technologies in favor of something far more participatory, particular, and open. Might it be possible, then, to take geo-visualization platforms such as Google Earth and Google Maps, which standardize and naturalize certain spatial imaginaries, perspectives, and coordinate systems, and pry them apart, creating fissures that open up spaces for multiplicity, otherness, and alternative worlds? We have struggled with this challenge throughout the life of the HyperCities project. HyperCities is software built on a dominant industry platform (Google Earth/Maps); it uses the satellite imagery licensed by Google; and marks-up data in a format (KML) that places objects on standard projections: Google Maps uses a variant of the Mercator projection, while Google Earth uses the World Geodetic System 84. WGS:84 is the reference system developed by the US Department of Defense for the Global Positioning System. Is it true that the master's tools can never dismantle the master's house, or is it possible to de-stabilize the world-views, epistemologies, and imagery comprising Google's digital globe in order to foster spaces of difference and particularity?

Unlike the objectivist and decidedly imperialist goal of accurately, comprehensively, and totally mapping every point on earth, HyperCities begins with a question: how can this platform be used to richly contextualize digital information, preserve individual memories, and, perhaps most ambitiously, begin to undo historical erasures and silences? Might the very technologies themselves—indebted to militaristic technologies of surveillance and precision guided destruction—be repurposed, as it were, for something life affirming? To do so, Google Maps/Earth must not be thought of as the mirror of nature, accurately reflecting the truth of the world "out there," but rather understood as a culturally and historically specific spatial practice that not only has its own logic and ideologies but also a particular place in the history of cartography, coordinate systems, and the emergence of the geospatial web.

(DS) Although they may have this effect, Google Maps and Google Earth were not, strictly speaking, intended to mirror nature; their construction reflects a series of compromises made for business purposes and strategic growth. Google Maps began as a geospatial search engine, and, hence, one of its most common uses is to find driving directions and business locations. Its use of the Mercator projection attracts criticism, as evidenced by indignant posts on the Google Earth Product forums such as: "Why does Google maps use the inaccurate, ancient, and distorted Mercator Projection?" Responding, Google employee Joel Headley explained that the Mercator projection best reflects the experience of the human viewer on the ground, who sees the world as a plane. Other projections, like Winkel Tripel, reflect the relative sizes of land masses more accurately at a global scale, but distort angles the further the user zooms in, so intersections and streets appear skewed. After careful consideration, including the release of a version that used a different projection, the design team settled on Mercator because they anticipated their customers would be using the maps at high zoom levels (i.e., "hyper-local") more than at a global perspective. The team had to choose between distortion at the global level and distortion at the local level, and the former was chosen for business reasons.[59]

(TP) Unlike most mapping applications, HyperCities was not designed to model urban environments, solve a technical problem, or provide data for businesses or driving directions. Instead, it was inspired by a humanities problem, namely how to conceive of history spatially, or, more precisely, how to give voice to historical erasures, preserve memories, and tell stories that move through time and space. While the past sometimes seeps out of the ground and buried memories sometimes come to light, more often than not, cities are places of erasure and oblivion. HyperCities is fundamentally a project of memory, one that was inspired as a response to destruction, even the very kind of destruction made possible by the technologies out of which it is built. In much the same way that Walter Benjamin's urban flâneur moves through a city but is "conducted downward in time," HyperCities users move along, annotate, and disrupt the digital streets while simultaneously peeling back and adding layers of "vanished time" (AP, 419). With tens of thousands of stories, photographs, data points, narrative collections, maps, and models, HyperCities enables the collaborative creation of "thick mappings" that are not only historical and commemorative but also speculative, strategic, and political.

The project does not presume to represent the fullness of any urban space as it once was through historical re-creation or mimetic modeling, nor does it seek to accurately reflect the reality of the city in digital space. Instead, HyperCities foregrounds critical, historical, and imaginative cartographic representations of the city by enabling users to interrogate, annotate, and remap the representations themselves. At the same time that maps bear witness to certain epistemological, social, and political configurations, they are also testaments to their own silences, exclusions, and erasures. While the maps in HyperCities have been georeferenced and placed within a GIS-based system that displays objects in standard global projections, they are not static representations or accurate reflections of a past reality; instead, the historical maps function as stacked representations in which one representation is keyed to another representation—not to reality. Digital mapping projects not only have the potential to destabilize and de-ontologize projections, coordinate systems, and representational cartographies, but they also have the possibility of enabling new modes of interactivity through memory maps, consciousness raising, and forms of counter-mapping that create fissures, alternatives, and sites of epistemological tension.

Georeferencing: "It is turtles all the way down"

(TP) The standard definition of georeferencing comes from ESRI, the provider of the ArcMap software, which operates on the standard ArcGIS desktop, and is used for making GIS maps and analyzing data:

> Aligning geographic data to a known coordinate system so it can be viewed, queried, and analyzed with other geographic data. Georeferencing may involve shifting, rotating, scaling, skewing, and in some cases warping, rubber sheeting, or orthorectifying the data.[60]

Another definition of georeferencing (at least for our purposes) comes from the anthropologist Clifford Geertz:

> There is an Indian story —at least I heard it as an Indian story— about an Englishman who, having been told that the world rested on a platform which rested on the back of an elephant which rested in turn on the back of a turtle, asked (perhaps he was an ethnographer; it is the way they behave), what did the turtle rest on? Another turtle. And that turtle? "Ah, Sahib, after that it is turtles all the way down."[61]

On the face of it, the two definitions are incompatible: the first strives to align a set of geographic data, such as a historical map, with a known coordinate system ("the basemap"). The basemap is considered the referent, and the other map or data is "matched up" to it. The process involves levels of compromise and distortion, from minor shifts to outright warping. In the end, the epistemology of the basemap triumphs. The second definition is a story of infinite regress, in which absolutes and origins are never found: it is turtles all the way down. Applying this logic to georeferencing means that we are referencing one turtle to another turtle, or one representation to another representation. Every representation, like every coordinate system, is made up. It's a way of representing a world that may or may not have any external referent. Aligning one representation with another representation does not yield a truth, but produces a relation, which allows a series of questions to be asked and various kinds of analyses to be undertaken. As long as the "basemap" is considered a representation and not a reflection of reality, the two definitions are perfectly compatible. But as soon as one representation ("the basemap") takes priority as the normative epistemology and everything else is measured according to its degree of deviation from this norm, we leave the world of turtles and enter the violent world of objective truths.

When we talk about georeferencing historical maps today, we usually mean making older maps line up with Google satellite imagery so that we can show them in Google Maps. The earth in Google Maps is a square Mercator Mobius strip developed through centuries of research into cartography that culminated in the development of satellites, GIS, and GPS. But this "basemap" is fundamentally a representation, the product of certain spatial practices, epistemologies, and compromises in seeing the world. The montage of satellite images is a series of choices such as using the Mercator projection to privilege the local over the global; it cuts off the globe at the eighty-fifth parallel in both directions and renders a rectangle rather than a sphere, given the limitations of the browsers available when the product was released.

If we look at earlier maps, we can understand the compromises that Google made and the ways in which prior map-makers saw the world. This history is not a march toward ever-more accurate maps but rather a series of contingent choices that made sense at different times for different purposes and, therefore, gave rise to varying world-views,

(14) Map of Ōmi province georectified using only a few control points.

epistemologies, and modes of representing. We might ask: what if Google Maps was not our "basemap" or reference point? What if we consider past geospatial information systems as perfectly "correct" ways of seeing and think of georeferencing as a process of translating between different discourses, of attempting to key one representation to another representation, and that this process of alignment can utterly fail when epistemological incommensurabilities present themselves? To do so, we will give a few examples of "spatial deformance," derived from the experimental interpretative

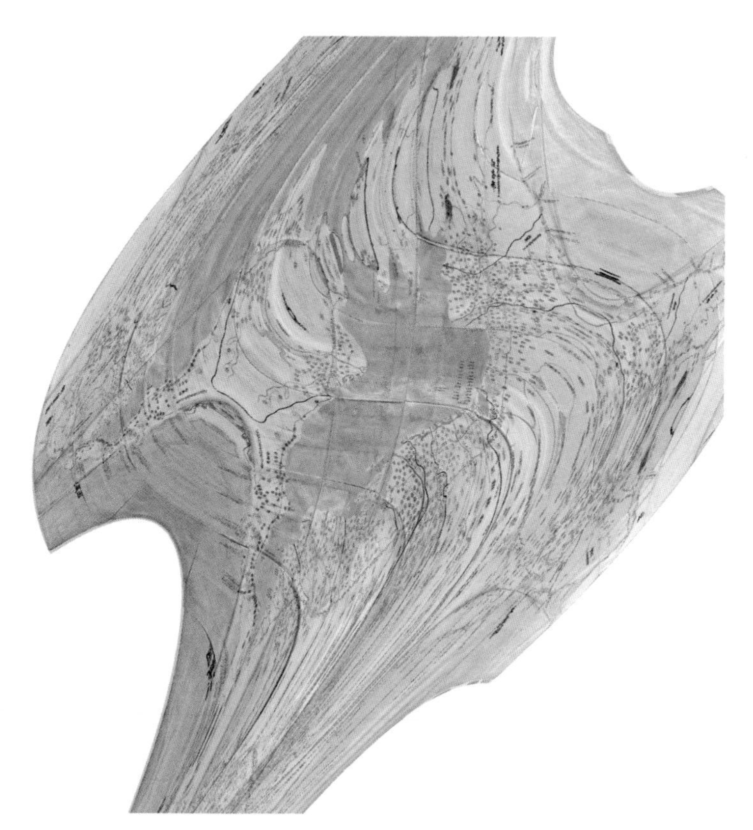

(15) Ōmi map georectified using sixteen control points and third-order polynomials.

and speculative process developed by Jerome McGann to analyze texts in *Radiant Textuality*. While McGann focuses on "poetic deformations" (such as reordering, isolating, altering, and adding), we focus on "spatial deformations" of the epistemologies built into maps.[62]

This map of Ōmi province in Japan is one of the provincial maps (*kuniezu*) commissioned by the fifth Shogun Tsunayoshi in 1696–1702.[63] (Fig. 14) These maps were based on cadastral surveys to gain a better assessment of the land, the people, local wealth, strategic points, and productivity immediately after the shogunate seized control over the realm. The map shows features of the natural environment, such as mountains and rivers, as well as the built environment, such as roads, villages, and egg-shaped ovals of varying size and color. An "egg" connotes the yield of rice from each village and thus signifies land value, not necessarily geographic space. Villages in the same province are designated with a uniform colored egg. The distance to major post stations and cities is indicated along the edges of the map.

The map is a geographic information system as much as a work of political symbolism. Navigation is difficult with this map because it is meant to be read from each of its four sides and thus lacks a single point perspective. Features and words on the map face different directions and even plotting a simple path from northern Nara to southern Yoshino is challenging. While Yoshino is located in the southern part of the province, everything in the region of Yoshino on the map (villages names, labels, images of mountains and waterfalls) faces southward, forcing the viewer to turn the map about 180 degrees. Moreover, the map does not represent geographic space and distances in a uniform way.

This map has been partially georectified to Google Maps using a few control points to position it roughly at the location it represents. We did this simply to display it in HyperCities and show an approximation of the area it represents. But more systematic attempts to georectify it warp the map into a smear. With sixteen control points, ArcMap produces a spatial deformance. (Fig. 15)

While the lake's geography lines up, perhaps surprisingly, with Google Maps, the distances between villages do not, let alone the "egg" sizes of each village. But this map of Ōmi certainly does not have to be read or judged by its lack of "alignment" with Google's satellite imagery and coordinate

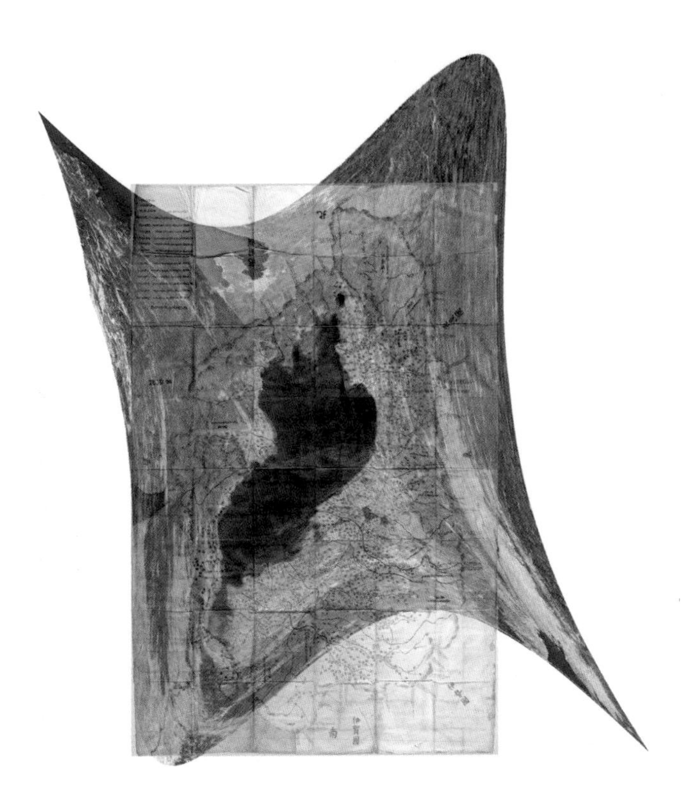

(16)

system; rather, it offers a spatial epistemology, a frame of reference for seeing a world, which was useful and accurate as far as the shogunate was concerned. If we take the map's epistemology as normative, we can georectify the projection used by Google Maps to yield another deformance. But because Google does not allow their imagery or maps to be reproduced as warped or altered in any way, we used satellite imagery provided by ESRI, which uses the same providers as Google (DigitalGlobe, GeoEye). (Fig. 16) This deformance was created using the map of Ōmi with satellite imagery superimposed on top of it through a third order transformation. Rather than georeferencing the Ōmi map to the satellite imagery, the satellite imagery was georeferenced to the Ōmi

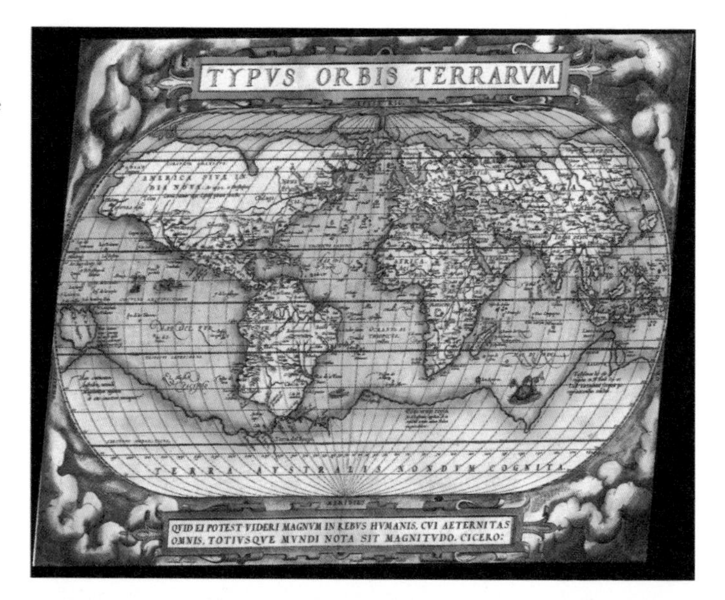

(17) Ortelius map, loosely geo-rectified to Google satellite imagery with four control points.

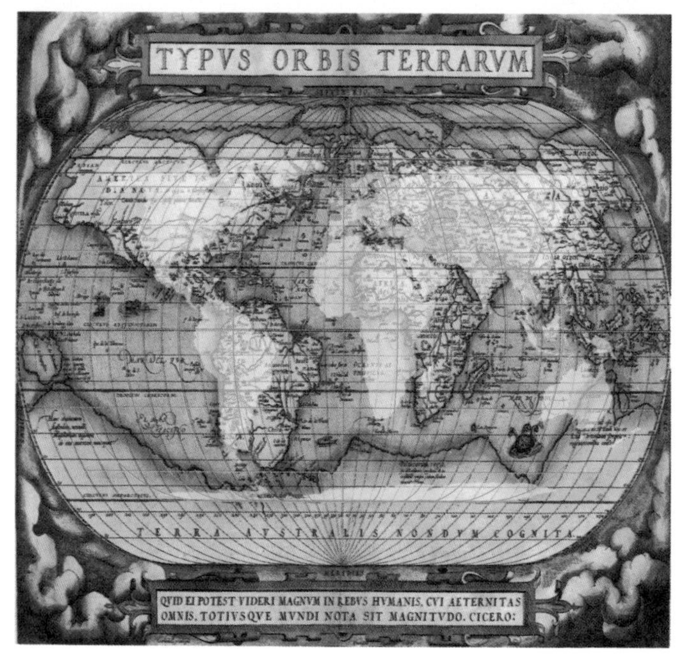

(18)

(19)

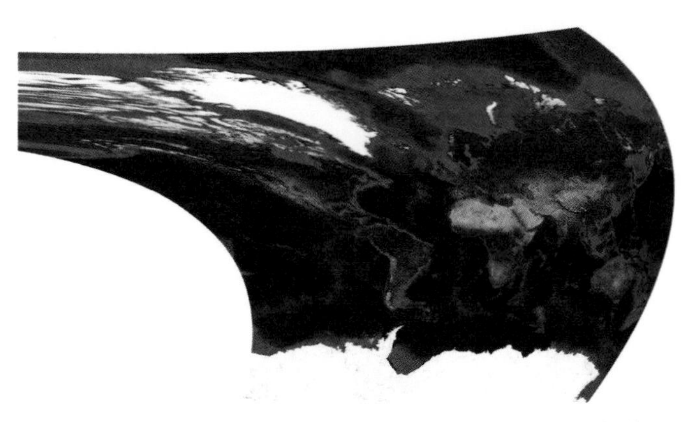

map. In order to maintain the shape of the lake, a total of twenty control points were used, distributed along the lake shore. Upon georectification, the imagery is stretched and warped to bring the control points into alignment.

More than a hundred years earlier, the Flemish cartographer Abraham Ortelius published a world map called *Typus Orbus Terrarum* (1564). (Fig. 17) Using four control points, it seems to roughly line up with Google Maps (despite the shapes of North and South America and the absence of Australia). As accurate as the shape of the European and African continents seems to be at this global view, if we overlay a grid of the Eckert IV spherical projection, subtler distortions emerge. The overlaid shapes of the continents show where they should appear in the projection: the map's scale is "wrong" in a number of places. (Fig. 18) Ortelius's map shifts all coordinates 30 degrees to the west (the distance from California to central Missouri). Trying to make his map truly conform to Google Maps leads to gross distortions.

We can point to his lack of data about the Americas as a shortcoming of his map, but what is more surprising is that Ortelius literally has a different point of reference. But, again, let us imagine the opposite: not a "deformance" of Ortelius's map but rather of the Mercator projection and satellite imagery that Google uses. In other words, let's georectify Google using the Ortelius map as our base-map. When we do so, the Americas are markedly distorted and the map pulls in toward the bottom and flares out toward the top. Part of this transformation is due to the fact that the Mercator projection is rectangular (Google uses a variant of the Mercator without the poles) and Ortelius's map is spherical; it also shows how much Ortelius overestimates the size of North America (which, at the time, was, like South America, largely unknown to him). But it also shows just how much Google's Mercator projection distorts: Google does not even try to retain a shape near to the Earth's actual shape, whereas Ortelius actually made an effort to represent the ellipsoidal shape of the globe on planar paper. When we try to re-project Google Maps' world to a different shape, it warps and the illusion breaks. (Fig. 19)

Ideologies of Accuracy

(TP) As much as "thick mapping" is interested in denaturalizing any geographic information system as a construction of projections, coordinate systems, and representational practices, georeferencing can also reveal ideologies at the level of spatial systems. This map was produced by the German Democratic Republic (East Germany) in 1970 to show the capital of Berlin. (Fig. 20) The map was published by the VEB Landkartenverlag in East Berlin, a company that existed from 1954 to 1976, and was considered the most important cartographic agency in East Germany. It was printed by the Military Cartographic Service based in Halle, Germany. Maps published in East Germany needed the approval of the Ministry of Culture and the Ministry of the Interior, both of

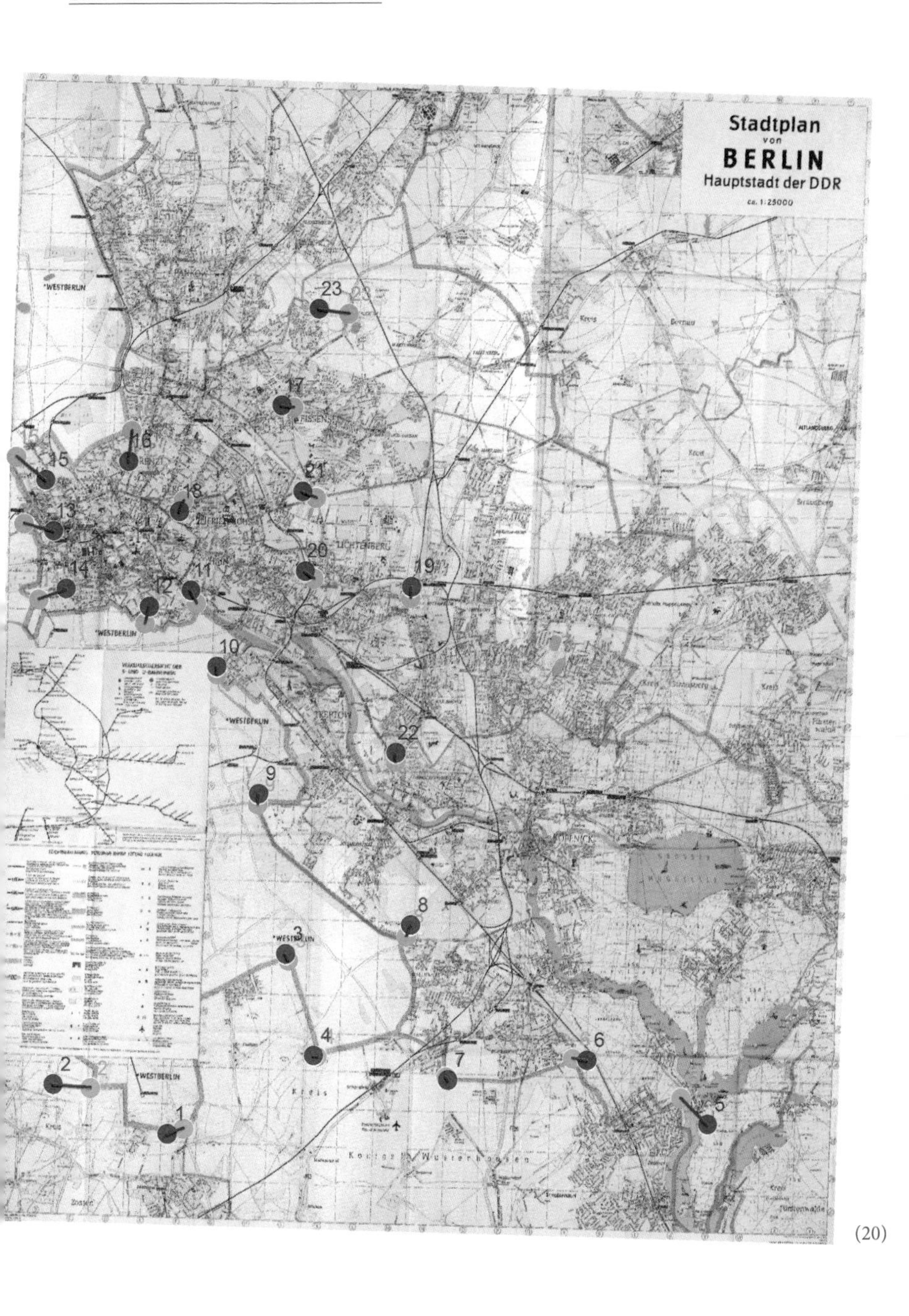
Stadtplan
von
BERLIN
Hauptstadt der DDR
ca. 1:25000
WESTBERLIN
WESTBERLIN
WESTBERLIN
WESTBERLIN
WESTBERLIN
LICHTENBERG
KÖPENICK
1
2
3
4
5
6
7
8
9
10
11
12
13
14
15
16
17
18
19
20
21
22
23

(20)

which imposed tremendous censorship and regulation. The map is striking, first of all, for the obvious fact that "Berlin" barely includes West Berlin at all. With the exception of a couple of major (unmarked) roads, green spaces, and subway lines that pass through the West, West Berlin is an empty land. East Berlin, on the other hand, is fully annotated with

---

Zero Hour: January 1, 1970

(DS)

```
CREATE FUNCTION dateLarger (dateOneIsBC tinyint(1), dateOne
datetime, dateTwoIsBC tinyint(1), dateTwo datetime)
BEGIN
DECLARE isLarger  tinyint(1);
DECLARE secondsOne  int;
DECLARE secondsTwo  int;
IF ((dateOneIsBC = 1) AND (dateTwoIsBC = 0)) THEN
  SET isLarger = 0;
ELSEIF ((dateOneIsBC = 0) AND (dateTwoIsBC = 1)) THEN
  SET isLarger = 1;
ELSEIF ((dateOneIsBC = 1) AND (dateTwoIsBC = 1)) THEN
  IF (YEAR(dateOne) < YEAR(dateTwo)) THEN
    SET isLarger = 1;
  ELSEIF (YEAR(dateOne) > YEAR(dateTwo)) THEN
    SET isLarger = 0;
  ELSEIF (DAYOFYEAR(dateOne) < DAYOFYEAR (dateTwo)) THEN
    SET isLarger = 0;
  ELSEIF (DAYOFYEAR(dateOne) > DAYOFYEAR(dateTwo)) THEN
    SET isLarger = 1;
  ELSE
    SET secondsOne = HOUR(dateOne) * 3600 + MINUTE(dateOne) * 60 +
SECOND(dateOne);
    SET secondsTwo = HOUR(dateTwo) * 3600 + MINUTE(dateTwo) * 60 +
SECOND(dateTwo);
```

streets, iconography, and symbols indicating major monuments celebrating the communist regime.

Georeferencing this map reveals another level of ideology, one that is not immediately legible in the cartographic symbolic systems. While the interior spaces of East Berlin are comparatively accurate when measured against Google's sat-

```
    IF (secondsOne <= secondsTwo) THEN
      SET isLarger = 0;
ELSE
      SET isLarger = 1;
    END IF;
  END IF;
ELSE
  IF (YEAR(dateOne) < YEAR(dateTwo)) THEN
    SET isLarger = 0;
  ELSEIF (YEAR(dateOne) > YEAR(dateTwo)) THEN
    SET isLarger = 1;
  ELSEIF (DAYOFYEAR(dateOne) < DAYOFYEAR(dateTwo)) THEN
    SET isLarger = 0;
  ELSEIF (DAYOFYEAR(dateOne) > DAYOFYEAR(dateTwo)) THEN
    SET isLarger = 1;
  ELSE
    SET secondsOne = HOUR(dateOne) * 3600 + MINUTE(dateOne) * 60 +
SECOND(dateOne);
    SET secondsTwo = HOUR(dateTwo) * 3600 + MINUTE(dateTwo) * 60 +
SECOND(dateTwo);
    IF (secondsOne <= secondsTwo) THEN
      SET isLarger = 0;
    ELSE
      SET isLarger = 1;
    END IF;
  END IF;
END IF;
RETURN(isLarger);
END
```

ellite imagery, the location of the Berlin Wall is consistently off by hundreds of meters in areas near the Brandenburg Gate and the region around Checkpoint Charlie and nearby border crossings. Although there is little visible distortion of the streets or even the river Spree, the map "bloats" the size of East Berlin in the interior and significantly expands East Ber-

- -

Documentation: a description of how it should work.

(DS) We take comfort, if not pleasure, in the dualistic opposition between the developmental sins of bugs and the virtues of features. The stranger world of undocumented features is a moral gray area between the two. MySQL's documentation guarantees that DATETIME fields hold dates back to January 01, 1000, at 00:00:00 midnight. Ask it to store the date of the Imperial Procession of 404 CE, however, and it willingly complies. It turns out that any date back to the first second of the Common Era can fit comfortably in the DATETIME field.

So why does the MySQL documentation not guarantee they will support dates beginning one second after 0 CE? Because they cannot guarantee it. The topsy-turvy world of date math requires that dates be stored as the number of seconds since a given period—today, that's January 1, 1970 at midnight. Good enough for business data that extends back maybe to 1900, but the majority of human history happened in the hundreds of billions of negative seconds.

HyperCities extends into the negative billions (10,000 BCE, the earliest date HyperCities can store, is 315,537,703,000 seconds BCE). A negative year in a DATETIME

lin along this contested region of the Wall. In other words, East Berlin is presented as "larger" than it really was, encroaching—quite consistently— hundreds of meters into West Berlin as indicated by control points 13, 14, and 15. Only through georeferencing this map is it possible to appreciate the ideology operating at the level of the projection itself.

---

field results in the unfriendliest of dates, 0000-00-00 00:00:00. Our solution: a flag indicating whether the date is a BCE date. How to query on two fields at once? The solution: dateLarger, the stored procedure reproduced above that I still do not quite understand.

What is the meaning of dateLarger? Another thousand lines of code. Writing complex applications requires converting table-based relational database records into objects using a piece of code called an Object-Relational Mapper (ORM). Friends in software development tell me to use one of the many already-available ORMs, for writing your own is the way of madness. After doing it myself, I agree. But most ORMs assume that we never use stored procedures to do comparisons. All but HyperCities' simplest queries use dateLarger, so ORMs are abstractions too leaky for HyperCities. Hence my still-growing thousand lines of code.

Of course, MySQL's DATETIME is designed in a Gregorian era for a Gregorian era. The date of 404 we assign is not the actual date Honorius processed. And in our database, the procession began at midnight, as good a time as any, for precision is relentless: to be precise in the present, we must be precise in the past. MySQL cannot allow fuzziness, but its precision leads to more fuzziness.

Of course, all maps encode ideologies. Ultimately, what counts as "accuracy" is a function of the goals of the cartographic representation: correct rice yields, planimetric precision, or keeping the citizenry of East Berlin in East Berlin? Google Maps makes choices about what counts for accuracy: making a map using a mathematical coordinate system involves assumptions and omissions as much as making a map where representing precise spatial relations is less important than showing rice yields. The process of georeferencing is the process of selecting one turtle, saying "this is it," and placing the other turtles on the back of the first. But the turtles never really line up, and, hence, we are always in the realm of productive deformance.

So what's at stake in using the projections and coordinate systems deployed by Google Maps/Earth? For HyperCities, there are strategic and pragmatic reasons to understand—and also to be able to deconstruct—the affordances, assumptions, and ideologies of any standardized cultural practice. The decolonization of knowledge may not be achieved by (ambitiously) creating indigenous humanities mapping platforms, but by strategically creating cracks and fissures in the most pervasive, ubiquitous knowledge platforms (such as Google Maps/Earth) and by dialectically tarrying with the negative of these platforms. Among other things, this means enabling an unbounded multiplicity of storytelling and counter-mapping that foregrounds contestation and alternative histories. In practice, it means the video testimonies of sixteen-year-old immigrant youth from LA's Historic Filipinotown telling their life stories on maps that can be opened simultaneously in scholarly narratives produced by tenured university professors, overlaid with decades of census data, and rendered on historical maps that show the history of redlining with the primary source reports created by the Home Owners' Loan Corporation in the late 1930s. These "thick maps" exist in a platform built on Google Maps/Earth, using the geographic information systems and satellite imagery of the imperial imaginary. Are they consumed by it? Do they inevitably speak its language? Or, perhaps, do these thick mappings betray the limits, contingencies, historical specificities, and exclusions of the dominant paradigm? The bigger question is how to interrogate the spaces for the production of what gets to count as knowledge at a given moment, the modalities for the production and ordering of discourse, and the conditions of possibility for the configuration of knowledge into systems, classification schemas, representations, maps, and other ordering environments.[64]

At its best, digital humanities helps to expose the data structures and classification systems, epistemologies and world-views, ordering systems and knowledge representations in ways that foreground their incommensurabilities. To be sure, we should imagine what it would mean to reconstruct an indigenous cartography which does not reference European geospatial systems or the ordering principles (planimetric, bird's eye perspective) upon which they rely. What, after all, happens to those spatial representations that cannot be georectified because they betray entirely incommensurable spatial systems rooted in different notions of proximity and distance, memory and community, duration and extension? Far from idle speculation, questions like this have formed the intellectual agenda of the HyperCities project and its approach to thick mapping. And, at the same time, we need to be able to deconstruct the spatial systems that have emerged as "standards" and "norms" by unmasking the world-views, relativizing the perspectives, and reanimating the voices that they left out.

There is much work still to be done in designing platforms that de-colonize knowledge and foreground epistemological incommensurabilities across cultures, languages, and historical periods rather than reinforce structuring hierarchies, standardized representational schemata, and exclusionary knowledge systems. We might entertain a series of future mappings that, for example, are not about the past or the present but are speculative, open-ended, and imaginary; or we might model mapping in the conditional or subjunctive tense, a notion that emphasizes what "might have been" or what "could be." In this regard, we might turn to the non-locative and ask: what happens to the speculative, the fictive, the imaginary, the non-real, the utopian, and the anarchic in the world of digital mapping? How might, for example, a situationist politic be re-appropriated as cultural critique? What would it mean to imagine a system of mapping using different projection and coordinate systems, of structuring relations based on models that were founded on radically different epistemologies and experiences of distance, duration, community, and relationality? What would such a "differential geography" look like and how might humanists design such a mapping platform?

On the other hand, what happens when we (willingly or not so willingly) participate in ever thicker and homogenizing social relations that enmesh real-time and real-space tracking to create a world in which the physical and the virtual are seamlessly connected in an augmented World Wide Web that resides in human bodies and physical things in space? What happens when the information web and the physical world are intertwined in a way that makes it impossible to de-link? Does the fully mapped earth radiate disaster triumphant? Might, for example, East German Stasi maps of every microscopic element of everyday life be among the "thickest maps" of all? In this sense, thickness is tantamount to authoritarianism. And what does it mean when this thickness is appropriated—on a technological scale with no precedent—by the democratic state, right here at home, through ubiquitous, inscrutable, and unchecked programs of dataveillance?

The challenge of inventing a "global cognitive mapping" that does justice to the place of the individual subject within ever-thickening systems of surveillance, mediation, networking, control, and global flows is still before us.[65] I think there are imaginative, possibly liberating or even weakly redemptive possibilities that could be invented and harnessed by "thick mapping" in the digital humanities, but these formations must be agile and strategic enough to perform their own deconstruction, to facilitate practices of mapping and counter-mapping, to interrogate the assumptions and aims built into any such practice. Such a deconstructive strategy is what Jean-François Lyotard once termed "paralogy" for its "imaginative invention" of giving rise to the unknown, disturbing the order of reason, producing dissent, and imagining a new move from within the order of things.[66] The imaginative ability "to make a new move or change the rules of the game" (52) by organizing and "arranging data in a new way" (51), for example, lies at the heart of curation or counter-mapping in a cultural-critical mode. It teases out sites of tension and possibility that give voice to particularity and expand notions of participation; it destabilizes and de-ontologizes representational cartographies, corporate platforms, and technologies—not to mention so-called social truths and publicly accepted norms—through new modes of interactivity, memory mapping, and consciousness raising. Thick mapping in the digital humanities is a cultural-critical praxis that engages not only with what is, but also with what might be, could be, and ought to be. It this sense, it is ultimately a kind of ethics, motivated by a responsibility to the other, a curatorial care for what is past, and an openness to what could be and might come. But, in the final and sobering analysis, there are no guarantees.

Rome: Jumping over the Line

According to myth, Romulus founded Rome by gouging a furrow to define the city's area. When his brother ignored the sanctity of this defining boundary, Romulus killed him. This story stands as a metaphor for the penalties imposed when one transgresses boundaries, whether topographic or disciplinary. At the same time, it emphasizes the unrelenting emphasis on lines as definers and describers of environments. In the twenty-first century, scholars regularly jump over disciplinary lines, and move beyond 2-dimensionality to consider kinesthetic and multidimensional interrogations of the historical record. Digital modeling capabilities, geo-browsers, remote sensing, geographic information systems, and accessible aggregated data sets promote thick mapping. From its inception the dynamic *HyperCities* platform has enabled the exploration of content as well as content delivery systems, encouraging scholars to take vertical, animated, collaborative, iterative leaps across and through fields, approaches, data, space and time.

The *Visualizing Statues in the Late Antique Roman Forum* project represented our first step in a continuing redesign of interface and argumentation in digital humanities scholarship. It uses the *HyperCities* interface to incorporate geotemporally referenced ancient inscriptions and texts, pictures, sculptures, reliefs, buildings, and urban spaces to interrogate the entry of Emperor Honorius to Rome in 404 CE. The NEH Digital Humanities Center Program supported research on this topic by Gregor Kalas (University of Tennessee) working with an interdisciplinary group at UCLA lead by Principal Investigators Diane Favro and Chris Johanson. The project team regularly met with Kalas in a recreated 3D real-time simulation of the late-antique Forum Romanum that operated as an idea-space. Moving together in the simulation the team collectively considered viewsheds, speeds of passage, statuary place-

http://inscriptions.etc.ucla.edu

ments and orientations, inscription legibility and repetition. The research process was an animated interaction that facilitated the blurring of boundaries between disciplinary domains and source materials. Kalas presented his final interpretation as both a static, textual narrative and as interactive text, embedded in the *HyperCities* viewing panel running alongside the interactive 3D models. As the readers progress through the text on the sidebar, the views of the models reposition to support the argument. Other sections of the web publication display information geotemporally, in a 2D map

with timeline, situating sculptures and inscriptions in time and space, directly linked to a comprehensive database.

The visitor to the *Visualizing Statues* website encounters a singular argument, and the 3D simulations and scholarly source materials on which it is based. The project invites readers to leap directly into the primary material and evaluate the author's assumptions and interpretations, thus blurring the distinction between hard publication and lab experimentation. In effect, the thick map has morphed into a thick environment that negotiates historical and contemporary actions simultaneously. The *Visualizing Statues* project helps to level the playing field between author and reader. The narrative, overlaid in a Google Earth environment, provides continuous spatial context. Unlike a traditional, two-dimensional argument, at no point does the narrative force readers to combine plan and section in order to situate themselves in the space currently under discussion. Moreover, the visual idea

space within which the author works is now exposed to the reader. Rome of the fifth century and its reconstructed monuments, the hypothetical locations of statuary, and the proposal for the imperial program of visuality in 404 CE are not imagined concepts, built from ekphrastic language, but instead are transformed into visual objects fit for study and interrogation in their own right.

Looking forward, it is not enough to fly from paragraph to paragraph like a digital, disembodied flâneur in a static city built of layered static maps. Always iterating, the research process breaks again across the line in lockstep with new technological affordances. Rather than fly above, we will soon drop down to ground level, to walk the cartographic streets as embodied avatars. Aiming to blur the distinction between map and experience, as we iterate forward we are combining gaming technology with the *HyperCities* blueprint to build an inhabitable world of maps and arguments. The sidebar interface of HyperCities remains, but a new narrative now runs inside a fully experiential, 3D world. The reader's view becomes more than a simple camera, but a digital approximation of embodiment. Visualizations of historic environments will vary, including both attempts at verism and experimental sketches of ideas—resembling maps more than reality. The viewer will simulate a walk through databases of maps, whose 3D buildings are extruded from the cartographic footprints.

HyperCities celebrates multi-vocality, but to date, only that of the authors, not the reader. Today multiple readers log in to read the *Visualizing Statues* project but without any knowledge of the reader community simultaneously engaging with the same argument. One future connects these readers. Users appear as avatars able to interact with others as they explore the arguments set out before them. Their paths will be recorded, later played back again as tours, comments, or critiques of the visual scholarship on display. The readers simultaneously author as they move through the space, but they also collaborate on live interpretations, reading aloud as they move, commenting as they go, activating spaces by their presence—from oratorical platforms, to ritual areas. The activity of the avatars builds a moving map, and takes steps toward creating a fully inhabited, living map. In essence through their simulated embodiment and interaction, readers walking side-by-side assess the visual and textual argument embedded directly in the virtual world. By turning off and on maps and emphasizing particular links for others in the world, they will transform the landscape. Theirs is now a shared experience of reading—a virtual journey, embedded in the thickest of maps, that requires virtual footsteps to move from point to point in a running narrative. This version of the embodied experience of argument, narrative, and evolved map is not yet built, but its foundations are in place. *RomeLab*, the 3-year study centered on Spectacle in the Roman Republic, has already developed some of the

http://romelab.etc.ucla.edu

experimental infrastructure.
A merger of this vision with that of the HyperCities code-base is on the horizon. As always, HyperCities has no truck with hypotheticals. Through rapid iteration, the process moves ever forward through a cycle of implementation, invention, and re-implementation. Why limit ourselves to talk alone, when we can continually jump across the line to make and to map?

Diane Favro and Chris Johanson

W
4

Mapping the 2009 Election Protests in Tehran

On June 16, 2009, Iranians took to the streets en masse in protests never before seen or experienced since the 1979 Islamic Revolution. The parents and grandparents who had ousted the Shah joined their children and grandchildren 30 years later to protest the rigged re-election of Mahmoud Ahmadinejad. Labeling his regime a "coup government," they held peaceful, silent demonstrations in contrast to what is by its very nature a violent act.

The first footage of thousands upon thousands of protesters, spanning the width of a number of grand boulevards across Tehran—Vali-ye-Asr, Enghelab, and Azadi—as far as the eye could see, sent chills down my spine. How could the Iranian government suppress this magnitude of discontent? Quite simple: by broadcasting the silent footage and claiming they are pro-government. Given that all media in Iran are state-controlled, the protesters cannot broadcast any counter arguments, so they changed their game: still protesting in silence, they used placards displaying rhyming slogans against the rigged election. The government played back by claiming the anti-government protests were taking place only in northern Tehran, an affluent area with Western educated and well-traveled citizens. (There is a strong class division between north and south Tehran, and the government's claims are easily accepted by the population in other provinces where they only know of this division in Tehran and base many assumptions on this fact.) From here, the protests gain momentum, and on June 20, they turn violent.

This is the point where my mapping project begins—an anger-infused and emotionally draining task that I saw as my duty to a homeland I barely know and only from childhood memories of summers spent with grandparents and cousins. My goal was to debunk the government's claims by showing that the protests are in fact widespread, happening in all provinces across Iran, regardless of class and ethnicity.

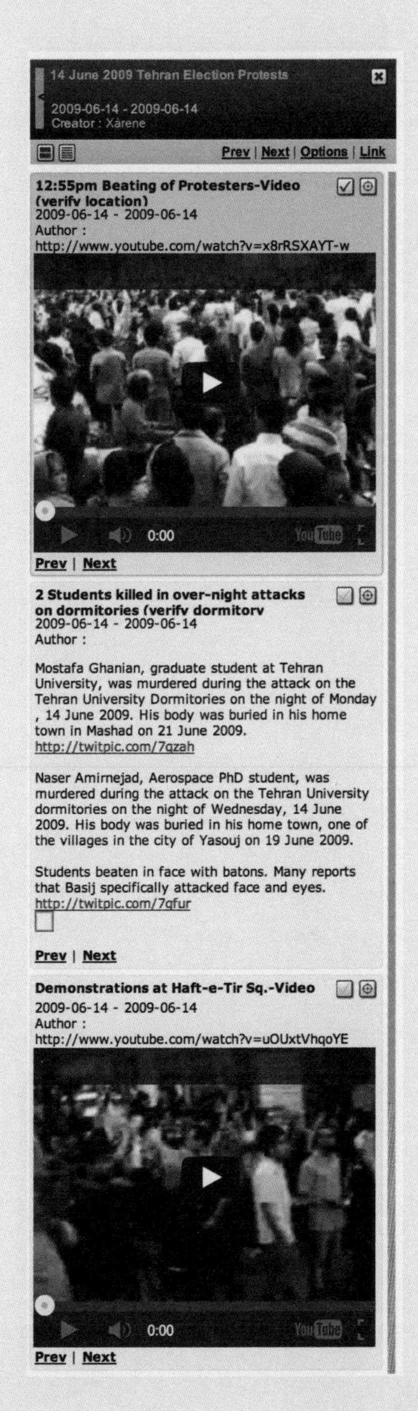

Tools and Methods

The protests across Iran were enthusiastically dubbed the first social media or "Twitter Revolution." Indeed, Twitter, YouTube, and Facebook played a major role in organization and communication among protesters, and between them and foreign observers; however, they fell short in creating a cohesive and easy-to-follow environment due to their single-event character, multi-temporality, and lack of a collective timeline. To bring the events together and to successfully disseminate information about them, I needed, first, accurate and time-stamped information of the events as they were unfolding; and second, a place to map the data.

Finding information was not difficult as the protesters immediately popularized a hashtag #IranElection with a staggering 221,774 tweets per hour. After hours of analysis and cross-referencing of tweets to other tweets and YouTube videos, I pinpointed the protesters giving accurate information on the events, and those

with a wider network across other provinces. In order to gain their trust—since government operatives caught on to Twitter and were using it to track protesters or to propagate misinformation—I would share new proxy information to bypass internet censorship, a highly sought-after barter-commodity in a Digital Black Market kind of way. (A Bay Area IT programmer, Austin Heap, consistently provided incredible proxy server support to protesters.) This allowed me to request specific types of information, primarily precise location and time-stamps as close as possible to the events.

For mapping the information, my first instinct was to use Google Maps, and by making the maps public I could send the information out to protest leaders and organizers across Iran. The goal was to boost morale in the face of censored news and deliberate media silence on the events despite obvious unrest in the cities. Because of the control of state media, the provinces are isolated from each other and from Tehran. For those people living outside of the province of Tehran, if they believed protests were not happening in the capital, they would easily abandon their efforts. Also, seeing the reach and occurrence of the protests in videos and images is far more powerful than reading and hearing scattered information online or through word of mouth. While Google Maps facilitated the mapping of links to city-by-city photographs and videos, the general layout and organization of the information was still visually hidden in links of titles to the windows holding the photographs and videos. At this point, HyperCities proved to be an ideal platform because all text and visual information could be mapped and displayed chronologically and visually—the windows holding the information could be opened to reveal all visual information at once, and by scrolling through the minutes of a single day, one could immediately get a sense of the events.

Emergent Information

Placing the information in visual and chronological order and analyzing it for accuracy of date and time led to an interesting study of protest slogans. I had been comparing slogans (chants and written signage) to make sure, for example, that the Quds Day protest videos were from 2009, not 2008. While fact-checking I noticed how the slogans shaped the momen-

tum of the masses, and also how media changed the slogans. What the protesters were chanting was not necessarily aimed at the coup government, but was a message for Western media, to show the true position of the people in regards to their government. Another facet to the language of opposition during this time was an increase of new poetry blogs, as well as a switch to poetry on existing blogs to avoid censorship and imprisonment. Iran has a terrible record of imprisoning and torturing bloggers (Hoder is still in prison; Omidreza Mirsayafi was killed in prison on March 19, 2009; and Sattar Beheshti was killed in prison around November 10, 2012).

Of all slogans, the one that has stayed with me, and surprisingly made the most impact since I made a specific entry for it on June 16, is this one:

You are the dust and dirt
Inferior and less than dirt
I am the light, I am the salt
I am the love-blind warrior
You are the blind force
You are the darkness
Brave and fearless I am
I am the owner of this soil

The slogan was aimed at the Supreme Leader's son and coup leader, Mojtaba Khamenei. From the point of view of the protesters the government is an oligarchy based on favoritism and familial ties, and no different than the hereditary system of the ousted monarchy. Interestingly, the wording also implies racial inferiority, being non-Persian. Whereas the monarchy consisted of Iranians celebrating Persian values, these Islamic leaders' loyalty is to an inferior culture—observant and mass-practiced and publicized Islam—and not in tune with the Islam of modern Iranians—secular and personal Islam intertwined with Persian traditions. It is important to note that the Islamic clerics and leaders have been trying to replace and even outright ban traditional Persian holidays, such as Chaharshanbe-Suri, which became a huge day of protest later in the year. Many other Iranian-centric slogans appeared and immediately became popular. This is significant because

for the first time since the Islamic Revolution, Iranians are also abstractly hinting at a separation of religion and state, something that can easily earn someone the title of "mohhareb," an enemy of Islam. It carries the death penalty. However, by November 4, the anniversary of the take-over of the US Embassy in Tehran, the protesters become more brazen, openly chanting: "Our race is Aryan, separate Religion and Politics!"

In all, the archive that I created mapped 22 critical days and more than 400 distinct events in Tehran between June 16, 2009, and February 10, 2010, using YouTube videos, Flickr photos, Twitter, Facebook, and other social media to create a digital record of the protests: screams, gunshots, beatings, clashes, blocked streets, sites of safety, and people killed. Many are very hard to watch, as they are maps of events, people, and voices no longer on the map.

Xárene Eskandar

Mapping Events/ Mapping Social Media

Participatory Digital Humanities

(TP) If there is a utopian idea at the core of the digital humanities, it would have to be the notion of "participation without condition"—that is to say, participation in the creation of the cultural record of humanity regardless of facticity and, hence, not dependent upon or restricted by race, gender, class, religion, ethnicity, nationality, age, language, access to technology, or education. Participation without condition means that anyone can tell his or her story, bear witness to events, and contribute to the collective record of humankind. This is both a retrospective and prospective record, historical and future-oriented at the same time, as the archive is not simply a question of what is past but, perhaps more urgently, a question of what is to come, a responsibility to a future that embraces, knows, and is informed by its many pasts.

Within public digital humanities scholarship, the concept of "participatory" has assumed critical importance insofar as it connotes the creation of conditions for engagement with communities and individuals not traditionally involved with humanities research and the documentation of culture.[67] Participatory is arrayed against the rules of prohibition and exclusion, the rarefaction principles and fellowships of discourse that Michel Foucault exposed in the creation of knowledge hierarchies and closed communities of practitioners.[68] In its best sense, participatory culture is open-ended, non-hierarchical, and trans-migratory, aimed at reestablishing contact with the non-philosophical, the heterogeneous, the people and perspectives left out, erased, and vanished. Participation without condition is not a principle that can be willed into place, but rather is an ideal to build toward through imaginative speculation and ethically informed engagement, one which promises—in the Derridean sense of the *arrivant*—to go beyond the limits and boundaries erected by prior formations of the humanities.[69] It is a motivating value rooted deep in the humanities, not a reality that can be built using technologies (although technologies can be enabling of new configurations for participation *as well as* their dialectical underbelly, new configurations for surveillance, control, and oppression).

Over the past decade, digital humanities scholarship has

Jan25 Voices @Jan25voices 30 Jan 11
Now: Tantawi is now Minister of Defence and Military Production. Live on Egyptian TV. Source: "This might trigger some heat." #Jan25 #Egypt
Expand

Jan25 Voices @Jan25voices 30 Jan 11
AlJaz : High rank army officer talking to protesters in Tahrir: they can stay. army will not interfere (as long as they stay put) #Jan25
Expand

Jan25 Voices @Jan25voices 30 Jan 11
@kringell @telecomix Please e-mail me about housing high B/W
View conversation

Jan25 Voices @Jan25voices 30 Jan 11
Tahrir: "Nothing like what they are showing on TV" Egyptian woman challenges Satellite News, minutes ago. http://tinyurl.com/62zgvw7 #Jan25
Expand

Jan25 Voices @Jan25voices 30 Jan 11
Audio: low-pass over Tahrir: Saturday evening audio from the square. Described as "cheering on helicopters". http://tinyurl.com/6l7msws
Expand

Jan25 Voices @Jan25voices 30 Jan 11
Conflicting voices: on the ground in Tahrir I have multiple calls saying calm, calling Arabya footage alarmist. #Jan25 #Egypt
Expand

Jan25 Voices @Jan25voices 30 Jan 11
Live audio: Just recorded a second call from Tahrir. Fighter jets and helicopters are loud. As is speculation as to their function. #Jan25
Expand

Jan25 Voices @Jan25voices 30 Jan 11
Audio re-post: "Mubarak Must Go" interview with young professional, the recording that seemed to crash twiturm. http://tinyurl.com/4pzzk9z
Expand

Jan25 Voices @Jan25voices 30 Jan 11
50,000 views crashed the music server I was using for audio. Now they are going up on my youtube channel: http://tinyurl.com/686sdjx
Expand

Jan25 Voices @Jan25voices 30 Jan 11
Those statements came from stated by Eryan (official spokesperson of Brotherhood). You may recall he was only released hours ago. #Jan25
Expand

begun to render the walls of the university porous by engaging with significantly broader publics in the design, creation, and dissemination of knowledge. By conceiving of scholarship in ways that foundationally involve community partners, cultural institutions, the private sector, non-profits, government agencies, and ever-broader slices of the general public, digital humanities expands both the notion of scholarship and the public sphere in order to create new sites and nodes of engagement, documentation, and collaboration. With such an expanded definition of scholarship, digital humanists are able to place questions of social justice and civic engagement, for example, front-and-center; they are able to revitalize the cultural record in ways that involve citizens in the academic enterprise and bring the academy into the expanded public sphere.

The result is a form of scholarship that is, by definition, translational and applied: it applies the knowledge and methods of the humanities to pose new questions, to design new possibilities, and to create citizen-scholars who value the complexity, ambiguity, and differences that comprise our cultural record as a species.

The "Voices of January 25th" and "Voices of February 17th" documentary projects are compelling examples of how

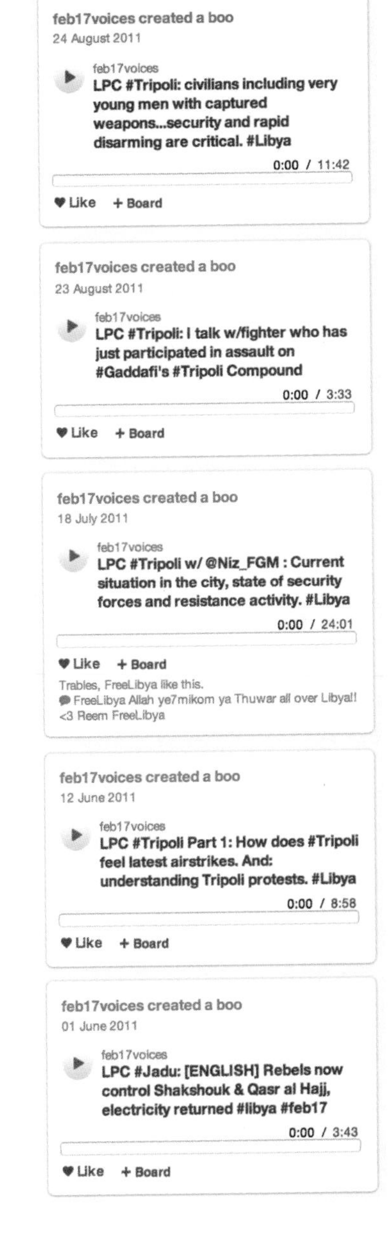

social technologies like Twitter can be used to give voice to people who were silenced in the 2011 revolutions in Egypt and Libya. Started by John Scott-Railton, then a graduate student at UCLA, the projects used Twitter to disseminate suppressed messages from dissidents to the world. Scott-Railton began the Voices of January 25th project when Egypt effectively "turned off" the Internet between January 28 and February 2, 2011.

Relying on cell phones and later landlines, Scott-Railton began calling friends in Egypt who knew protesters and could provide highly localized, accurate accounts of what was happening on the ground. He assembled a network of trusted informants who agreed to have their phone calls recorded and their voices published to the world on AudioBoo, an audio hosting service. Scott-Railton, then, posted messages to Twitter, often with links to audio files and other media reports that would help the world "see" and "hear" what was going on in Egypt in real-time. In effect, the digital portal became a global public sphere, however fragile and endangered, that was fundamentally linked to the deeply embodied and precisely located events on the ground. The voices are now part of a living web archive and documentary memorial.[70]

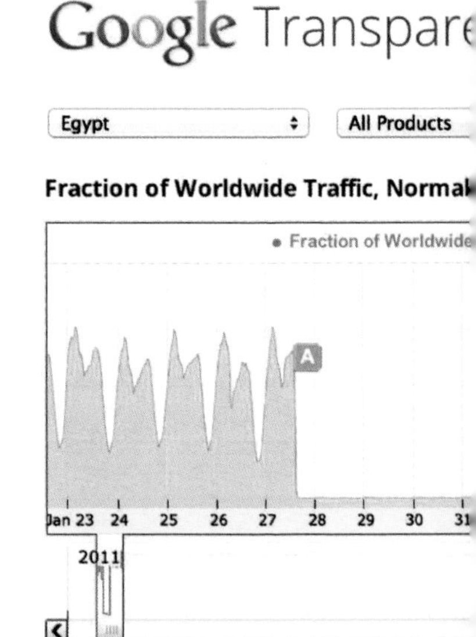

While the role of social media has been feverishly debated in fomenting, planning, and sustaining revolutions since Twitter was first hailed—somewhat exaggeratedly—as a revolutionary technology in Moldova in 2009 and YouTube became a people's archive for election protests in Tehran during the summer of that same year,[71] it seems incontestable that the "public images" produced by broadcast media (often singular, uni-directional, and hierarchical) are being supplanted by decentralized, multi-directional "public utterances" that are changing the way in which events unfold, become represented, and are disseminated (almost instantaneously) on a global scale. Two decades ago, Paul Virilio thought that the physical space of the public sphere had become replaced by the media of the "public image"; but now we are seeing the reassertion of the public sphere through radically non-Cartesian geographies enabled by what I earlier termed the web's "contiguity of the non-contiguous."[72]

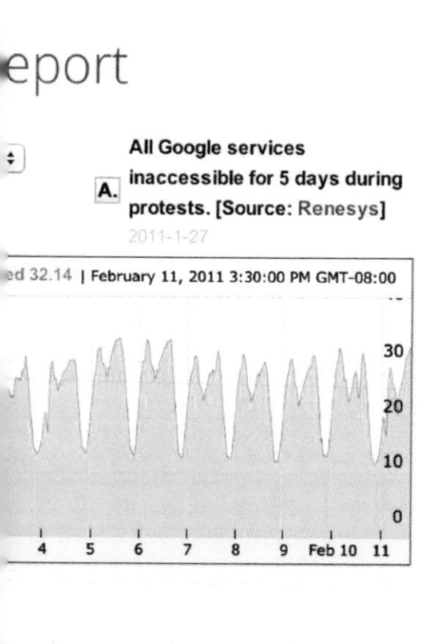

What this means is a massive contraction and alignment of the event (an embodied and location-specific phenomenon), the representation of the event (through Twitter messages, blogs, video, and photographs captured and uploaded from mobile phones, and so forth), and the dissemination of the representation (through web-based social networks and information channels). But more than just an alignment of event, representations, and dissemination, we are also seeing complex feedback loops in which the representations and disseminations instantaneously influence the course of events. As such, the event is no longer the same event because it is constantly re-constituted by other events, representations, and disseminations that recursively coalesce. The result is a significantly more adaptable, amorphous, global, but also ephemeral, public sphere, one which may, for example, be constituted as a contiguous space connecting Westwood, California, and, simultaneously, Tahrir Square in Cairo or Benghazi, Libya.

At the same time, we need to be critical and suspicious of any evidentiary function of social media, not only because they can be easily manipulated and are hard (although not impossible) to verify, but also because social media can be used by anyone, in the service of both democratic and authoritarian ends. And even with the best intentions, social media can amplify misinformation on a global scale, creating an echo chamber of falsehoods that are easily accepted as truths by virtue of their sheer repetition. There is no clarity of meaning or channels of truth to be found here, only multiple levels of mediation within complex and ever-shifting dynamics of power and participation. We are just beginning to understand the transformative possibilities, which are imbricated in ideologies of control—but also possibilities of democratization—of these multi-mediated public spaces.

Around the same time that Scott-Railton began the Voices of January 25th project, the HyperCities team created a mash-up for live streaming and archiving Twitter feeds from Egypt and visualizing them on a Google Map. The project, "HyperCities Now," made live calls to the standard Twitter search API for tweets originating within 200 miles of Cairo's city center and containing hashtags such as #jan25 or #egypt. This feed then placed the tweets on a Google Map, displaying a new tweet every four seconds, and populating an endless time-bar, down to the granularity of a single second, across the top of the interface. Although not 100% accurate, the tweets' location is based on users' self-reported profile locations, or coordinates provided by their cell phones. Sensitive to the risks involved in creating an archive that precisely maps Twitter users, messages, and location, the HyperCities team truncated the exact latitude and longitude (when it was returned by Twitter's location parameter) at the hundredths decimal place, effectively placing a tweet within a 2 mile radius, rather than at an exact GPS location. Over the course of several weeks, about 450,000 tweets from Egypt were archived, with the most tweets (nearly 10,000) occurring in the hour in which Mubarak gave his resignation speech on February 11, 2011.

In all, the project archived tweets and retweets from about 40,000 distinct Twitter users documenting the events of the Egyptian Revolution. While this number is significant (and it will provide the basis of our analysis below), we need to bear in mind that only about 5% of Egyptians actually used social media during the revolution, and most Twitter users were part of a fairly homogeneous group in terms of education level, class, and generation.[73] It would be a mistake to equate the Twitter archive with the Egyptian revolution or suggest that it singularly catalyzed the revolution. As Ramesh Srinivasan has argued, tweets from the revolution tended to reflect the democratic ideals of the more educated protesters, while working class protesters in Tahrir Square (who were not on Twitter but formed the majority of protesters) tended to express more pragmatic and economic reasons for their dissent, such as skyrocketing food prices.[74]

Scott-Railton's "voices" project represents a wider (and arguably more reliable) cross-section of constituencies since the conversations were with different generations of protesters via cell phone and landlines and only published to Twitter once he verified the accuracy of the information. Unlike the algorithmically aggregated, displayed, and archived data of "HyperCities Now," Scott-Railton's work was possible because of ever-expanding, although deeply fragile, networks of human witnesses who trusted him to listen to, steward, and relay what they saw, heard, and experienced. In this sense, the project accords with an earlier project of curation created by Xárene Eskandar, also a graduate student at UCLA, documenting, day-by-day and often hour-by-hour and sometimes even minute-by-minute, the election protests in Tehran during the summer and fall of 2009. Utilizing the HyperCities platform, Eskandar painstakingly documented gunfire, protest sites, photographs, Twitter messages, and YouTube videos, creating a geo-chronology for hundreds of reports and media objects that she found online and through networks of contacts. This project, like Scott-Railton's, was profoundly connected to the original etymology of the term curation, meaning "care of souls" or, in some cases, "stewardship of the dead." Both sought to curate—care for, preserve, document, and archive—the lives, experiences, and actions of the protesters for a global audience, despite (or perhaps because of) the precarious material, social, and technical conditions of possibility for the very stories.

While HyperCities Now curates data in different ways, it has an analogous goal of expanding the concept of the public sphere through an ethic of amplifying participation. Each of these projects was deployed extremely quickly—in a matter of days—as experiments or prototypes to intervene in an event that was still unfolding and unbounded. Far from complete or total archives documenting "the whole history" of the revolution, they are motivated by several principles that I think accord with critical theory as a socially engaged praxis: a respect for multiplicity and difference through the creation of trusted social bonds, an approach to historical documentation that builds from the fragments of participatory discourse, and a concept of archivization made possible by the contingent material technologies of communication (ranging from mobile phones, social media applications, and decentralized data centers to MySQL, PHP, and JSON feeds). These archives are not simply documents of the past ("what was") but are spectral, in that they pose haunting questions about the possibility of a future and, therefore, are motivated by a responsibility or promise that remains open and undetermined. As much as we may hope for a coming democracy, the future may also bring the disaster, and this is something with which these projects hauntingly reckon.

In addition to Egypt, the HyperCities team mapped and archived Twitter feeds from Libya as well as Japan following the earthquake, tsunami, and nuclear disaster. In the case of Japan, precise location was important for coordinating disaster relief and remains a priority in other mapping projects undertaken by the team, including one called "Bishamon," which aims to precisely track airborne radiation contamination from the Fukushima power plant. Spearheaded by a team of volunteers from GISCorps and CrisisCommons, the project initially mapped hundreds of thousands of social media feeds onto GIS data (including flood zones, evacuation centers, traffic, and public phone locations) so that real-time decisions for coordinating relief efforts could be made. It has evolved in a number of directions through collaborations between UCLA's Institute for Digital Research and Education, led by Yoh Kawano and David Shepard, and other institutions, including Harvard University and Niigata University in Japan. With Harvard, the UCLA team provided extensive

http://beta.jdarchive.org/en/home

content for the "Digital Archive of Japan's 2011 Disasters," and with Niigata University, the team has developed a web interface and visualization tool for mapping real-time radiation levels. These projects are discussed more extensively below.

On the Event: HyperCities Now (Egypt, Libya, Japan)

Every event is singular but enmeshed in an infinite number of other events (space).

The event, its representation, and the dissemination of the representation coalesce in feedback loops in which the event is never the same.

Events happen on bodies and are expressed in language and through visual media.

Events are processes of becoming with multiple temporalities.

The event has no end (time).

(TP) A message flashes on the screen. It is by a user who goes by the handle PortSa3eedy: "Chants in Tahrir: 'Muslims, Christians, We are One.' This is the rebirth of a country, with a feeling that anything is possible #Egypt." The time is February 11, 2011, 07:22:55 PM in Egypt. It's about an hour after Mubarak delivered his resignation speech. The location returned is city-center Cairo. The Twitter message stays on the screen for about four seconds, before another message flashes on the screen. I watch for hours, as thousands of messages ap-

pear and disappear on the map, which shifts as the location changes. It is a mesmerizing experience, quite unlike that of watching television, where reporters "go on location" to interview, at most, a few dozen eyewitnesses and provide pre-digested commentary and analysis for the viewers' consumption.

Watching HyperCities Now, I have no idea what to expect from the people "speaking" through Twitter. It's a raw, live, and unpredictable cacophony of voices, expressing hope, joy, and unbounded openness to a future in which anything seems possible. There are thousands of such voices, each embodied in messages of 140 characters in length and posted to Twitter, most likely through web-enabled mobile phones. Many are voices of people in Tahrir Square at this very moment in time, reaching out to a public that far exceeds the apprehension of any human listener in this place, at this time. On Twitter, the singularity of the voice is transported, refracted, and multiplied. It becomes part of a collective event and attaches itself to anyone ready to listen, anyone attuned to its message, anyone standing open.

Every tweet that flashes up on the screen is a pure event. It cannot be predicted and, in a sense, breaks free of time; yet when it happens, the tweet is utterly bound to a "kernel of time" (*Zeitkern*). In the same way, the tweet cannot be specified precisely on a map ahead of time, but when it happens, it is utterly "incarnated," the action, thought, and message of a person in a particular place. As a viewer, I cannot know when a tweet will come, from where it will come, who will send it, and what it will say. In this sense, it is a flash, a moment of truth of someone sending a message into the ether.

Only after the event can we begin to make sense of it as an event, by putting it in a timeline, by considering it as part of a network, by analyzing it for meaning, by investigating it as a part of a whole. This is the realm of hermeneutics, computational analysis, statistics, historical interpretation and explication. It happens after the fact.

There are three kinds of temporality at work in HyperCities Now:

The first is the time of the flash, the Aion as pure event, the Twitter message that appears on the map at a singular moment in time. It is an absolute surprise, utterly unexpected and unpredictable, even though its conditions of possibility can be anticipated and its context understood. This is the "now-time," the flash Walter Benjamin called the *Jetztzeit*, an instant or eruption that distills or contracts the present and harbors the notion of the revolutionary, the messianic, the possibility of a future that cannot be derived from the stock

of historical experiences. Anything is possible: justice, democracy, freedom, but also the violent backlash, the disaster, the catastrophic.

The second kind of temporality is that harbored in the event as network, when it becomes an event with hopes and expectations of a future that is not yet and memories and traces of a past that is no longer. It is reflected in the viewer's perception of the flash itself: when we—those who are corporeally "uninvolved" in the event, those who are "far away" as spectators—perceive the Aion, the event has already happened (it is no longer) and another event is about to happen (it is not yet). In both, there is a deep contingency. Recalling Aristotle's definition, the contingent is that which is neither necessary nor impossible. It didn't have to happen this way, and it could have been otherwise.

The third kind of temporality is that of Chronos, the historical time in which events occur, the timeline of events within chronological time, which help us make sense of events for analysis, comparison, and hermeneutical interpretation. It is the archive of tweets, searchable by hour and day, viewable by minute and second, in relationship to every other tweet by any other user who also used certain hashtags to describe, document, or otherwise witness events on the ground, in Egypt, during the revolution that began on January 25, 2011.

Start-dates are comparatively straightforward to assign, especially when world-historical events have demarcated triggers, such as January 25, 2011, or March 11, 2011, the day of the earthquake, which struck at 2:46 PM, and tsunami that enveloped Japan shortly after (3:38 PM). Such events cannot be readily derived from prior history (although their meanings and understandings will inherit much from the stock of historical knowledge and experience). They could not have been predicted to occur *on that day, at that time*. For that reason, world-historical events are first of all events that cannot yet be comprehended. They don't yet have full meaning because they are eruptions, fissures that are violently torn from the everyday. For someone "inside" the event (by which I mean someone whose body is affected by the event, who is physically proximate to the event and is, therefore, touched by it in a way that is different from someone who experiences the event through mediations and representations alone), it is impossible to know the extension and duration of the event, its scope and scale, its impact and significance, its resonance and meaning. The event is happening, and only later, with the passage of time and the (comparative) clarity of distance, can eyewitnesses make sense of the event by perceiving its contours. For someone "outside" the event (whose body is ostensibly safe and untouched), the event assumes certain contours right away, even if these contours are erroneous or incomplete, through the media in which it is represented and disseminated (on television, on radio, on the web, via Twitter, and so forth). It is foremost a spectatorial event, which also has the effect of placing the viewer in a precarious ethical relationship to the suffering of those "inside" the event.

I watched the Twitter streams flash on the map from the safety of my desk in my office at UCLA, placing me in a questionable ethical position of a privileged observer of events that I took no part in. For the person sending a message, the vulnerabilities cannot be overstated: who is to know if the message will be tracked and put into another kind of database, not one that archives the voices of history but one that aims to squelch those very voices? Who is to know if data will be harvested by others in order to more fully exert control and homogenize history? I am only an ethical spectator—a witness to the witnesses—if I am willing to be open to the message or voice of the other, whether the call for help, the call for solidarity, or the wish to know, document, and remember. I am a witness to the witnesses for the sake of a future to come.

But when does an event "end"? Who would dare say, at any given moment, that it is now over, that people who were there can (or should) "get on with their lives"? The event has no end for anyone who has survived a catastrophe, who has lost their family, who has been exposed to radiation, or who has risked everything for a future democracy. And not only does the event not have a temporal end, but every event opens up onto an infinite number of other events—both smaller and larger, directly related and tangential—at any given moment. Every event is part of a network of infinitely many events, which are impossible to specify or document in their entirety. Events are ever thicker networks of events, no matter how zoomed in or zoomed out one is. There is neither an end, in a temporal sense, nor an end in a spatial sense, to the event. In this regard, Gilles Deleuze was quite right in his argument that every event is singular (Aion) but enmeshed in and emerging out of and into an infinite number of other events: "each event communicates with all others and they all form one and the same Event, an event of the Aion where they have an eternal truth."[75]

Chronos

(TP) The HyperCities team developed "HyperCities Now" over the course of three days in late January of 2011, as a way to visualize and archive social media messages sent through Twitter relating to the events unfolding in Egypt.

During the days that the Internet was shut down in Egypt, the data are very limited. We started archiving on January 30 and stopped on February 25, not because the event was "over" but because we kept crashing our hosting service (and received several requests to terminate all scripts that we were running on their servers). Moreover, we wanted to establish a reliable partnership with our Digital Library for metadata creation and data preservation as well as a process and methodology for archiving "social media/events." We had made the decision early on to archive only tweets with a location parameter (either provided by the user or from the user's device), for the sake of more likely archiving tweets from people who were there on the ground.

To be sure, this decision already precludes certain questions from being asked, such as how events in Egypt resonated on a global scale or in other national or regional contexts. The idea of focusing on tweets with a location parameter within a 200 mile radius of Cairo's city center with certain hashtags (#jan25, #tahrir, and #egypt) was to sift out some of the inevitable "noise" that could easily swallow up or mute the voices on the ground. In a little more than three weeks, the HyperCities Now team streamed and archived more than 450,000 tweets, mostly in Egyptian Arabic and English.

A few weeks later, the disasters struck Japan and the team immediately enabled the streaming and archiving of Twitter messages from the entire region around the epicenter of the earthquake, near Sendai, Japan. At first, Twitter feeds on maps functioned as real-time "reporting" for coordinating disaster relief, especially when social media data could be linked with other GIS data, such as flood zones, blocked roads, and functional cell phone towers. We began to analyze the Twitter feeds from Egypt and the Japanese disasters from the standpoint of a multiplicity of scales and within networks of different kind of relations. It would be (nearly) impossible, of course, for any human to read and react to all the tweets, as they are also embedded in a network of infinitely many other tweets, photographs, weblinks, and other documents attesting to events. The scale defies human comprehension, whether it's the 700 million total Twitter messages during the month after the earthquake and tsunami in Japan or even the significantly smaller sample of 660,000 messages archived by UCLA that were bound by geolocation and certain hashtags.

As such, the team focused on scaling data to answer various kinds of questions, essentially "toggling" between the singular voice or story of a particular eyewitness and global, aggregate experiences. The latter does not represent the reality of "Tahrir Square" or "the earthquake and tsunami" (as complete or total events) but rather the totality of our archive, and therefore, can only reveal structures, patterns, and visualizations of the archive's data. This is not entirely distinct from what historians already do, insofar as they select source materials and emplot events at various levels of "zoom" in order to convey different kinds of meaning. In other words, scholars of history frequently find themselves "toggling" back-and-forth between macro-level accounts of the event (zoomed out) and micro-level accounts of individual experiences (zoomed in), which are, by their very nature, defined by their specific experiences, perspectives, spectatorship, language, and so forth.

Zoomed out, computational analysis of "big data" allows us to visualize and interpret over-arching structures, relationships, and patterns, something that accords with the methodological and theoretical approaches that have begun to emerge in fields variously called "culturnomics" or "cultural analytics," to use the term coined by Lev Manovich to describe the quantitative analysis of large-scale cultural datasets.[76] In practice, it is similar to the work of "distant reading" described by literary scholar Franco Moretti as a practice that moves away from the close, hermeneutical reading of individual texts in favor of an algorithmic approach that aims to visualize and understand entire corpora through attention to systems, structures, patterns, and trends.[77]

Whole corpus analysis potentially facilitates the democratization of knowledge, as no subset of texts becomes canonical. Instead of privileging "human reading" (in which we necessarily have to limit ourselves to a tiny canon of works or a small selection of reports, documents, and voices), distant reading is performed by a computer and can easily "read" thousands, if not millions, of works. As such, "distance listening" to Twitter feeds from around the world might one day facilitate a democratization of witnessing, since it has a leveling effect in that all utterances are granted equal importance and weight, such that no one voice takes priority or assumes canonicity. This democratizing effect is, of course, always a product of the data records and the data structures of the database, which may be ultimately agnostic to content, individual experiences, and even individuals, but are still produced by human beings who make decisions about what data to capture, how to structure it, and how to make it useful.

(DS) In this regard, data are not preexisting, observable facts that supposedly exist "out there," independent of the observer, waiting to be visualized; instead, they are a function of the means by which they are captured and the interpretive decisions made. As Johanna Drucker has pointed out, all data are really capta, and visualizations should acknowledge the observer's role "using interpretations that arise in observer codependence, characterized by ambiguity and uncertainty, as the basis on which a representation is constructed."[78] Our archive of tweets might be better characterized as a "captaset": a collection of utterances by those who used Twitter to comment on (and perhaps, to a certain extent, influence) events, sampled from the data Twitter made publicly available through its API. We privileged some hashtags (#egypt, #jan25, and #tahrir) over others. And, of course, Twitter itself is a self-selecting community, primarily youth, many from educated and wealthier backgrounds, who had access to this technology.[79]

Egypt: A Global View

(DS) What allows us to call the Egyptian revolution an "event"? We cannot have faith in a definition of the event in the singular, a suspect concept once we begin to consider events in layered, multi-functional contexts. Instead, what if we imagine an event as a network, and try to visualize it as such? Networks invite us to imagine peers working together to pass on information, without official communication channels. Network analysis is an apt method for analyzing events represented and enacted through social media. Ideas (whose signifiers are hashtags) are simply nodes, manifested only by the individual acts of communication users send to each other. The diversity of discourse bears witness to the complexity of an event and the number of people and messages involved; in theory, networks show many revolutions rather than one revolution.

Figure 21 shows all the hashtags used in tweets gathered by "HyperCities Now" during the early period of the Egyptian revolution[80]; each node is a hashtag and a line between two nodes represents the messages that contained those two hashtags.[81] This is the global level of an event on Twitter. Nodes' size reflects how frequently they were mentioned. We recorded these tweets by tracking three hashtags: #jan25, #egypt, and #tahrir, but 3,520 other hashtags occur with

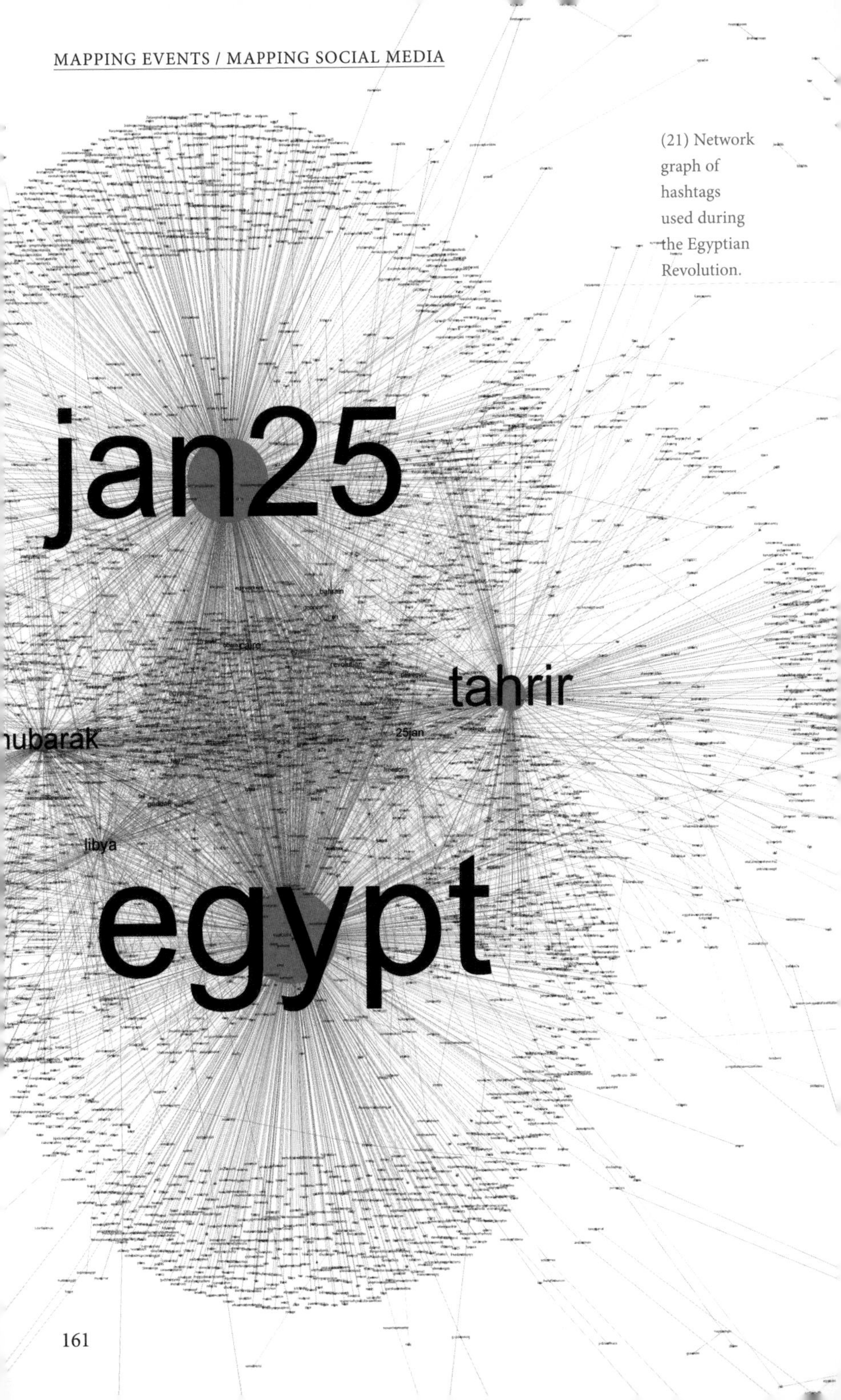

(21) Network graph of hashtags used during the Egyptian Revolution.

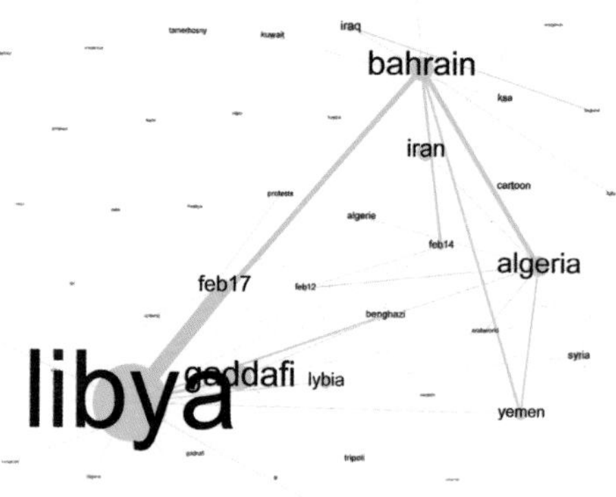

(22) Egypt's "Other Countries" community.

these three hashtags. This network looks like a galaxy with the three tags we tracked as the suns that exert their pull on the other stars.

Alongside its suns, this galaxy has solar systems. In network theory, these are called "communities," sets of nodes more strongly connected to each other than to other nodes.[82] Colors highlight different communities. The red nodes are those most strongly connected to the three central nodes. This is the "Egyptian revolution" community. Not surprisingly, it is the largest. There is another smaller one, centered on #mubarak. There are, in fact, 200

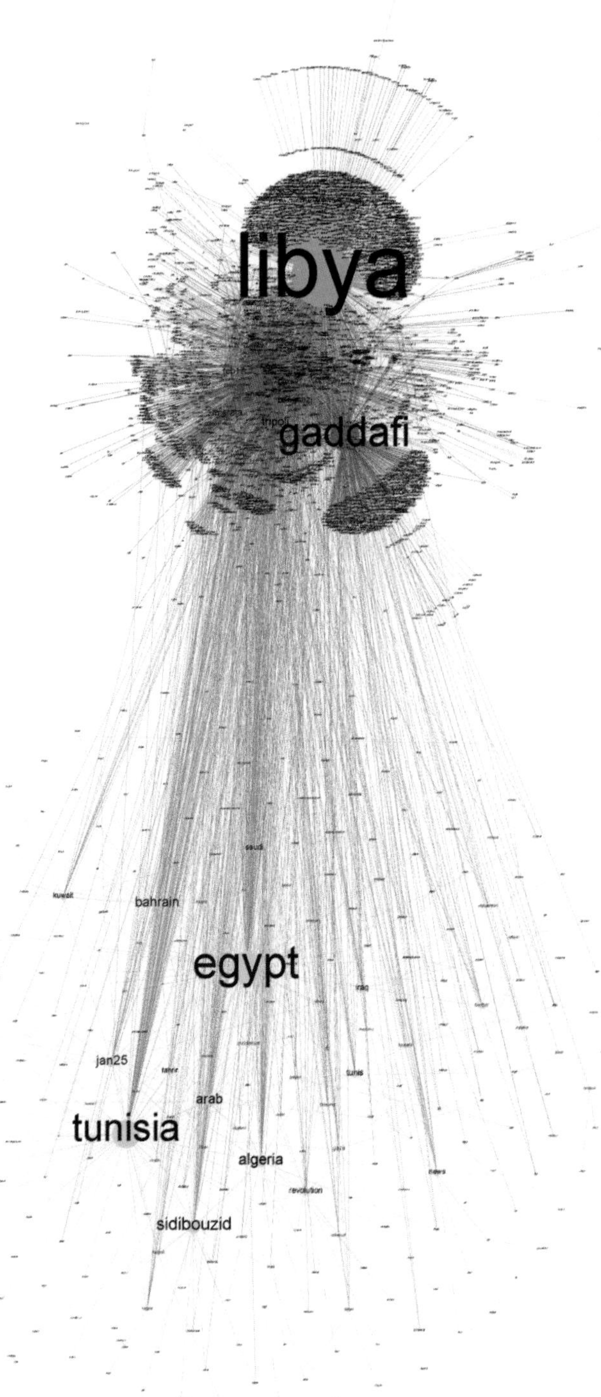

(23) Network graph of hashtags used during Libyan revolution.

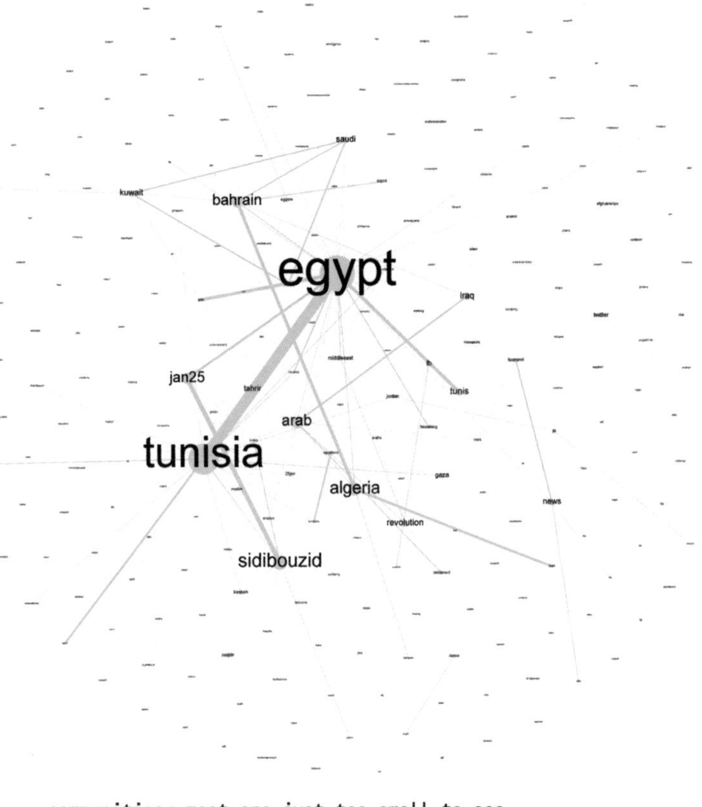

(24) Libya's "Other Countries" community.

communities; most are just too small to see.

These solar systems/communities have their own centers, which highlight the many centers in a galaxy of hashtags. Derrida argues in "Structure, Sign and Play" that the center is a function not an intrinsic unity, a site of origin, or a telos.[83] What if the "function" of our center is a math function? In this case, it is the Louvain method for community detection, which discovers latent patterns in networks.[84] It shows that networks of speakers self-organize their communications around particular topics.

In Figure 22, we see yet another community, "other

(25) Merged network graph of hashtags in Egyptian and Libyan revolutions.

countries." Its hashtags are mostly the names of other countries or key places where Arab Spring movements occurred. #libya is its strongest node, about 9% of the size of #egypt and #jan25. This community shows the discourse surrounding the Egyptian revolution opening to other events, which are related yet distinct. One event opens to others, showing the porousness of event boundaries, and yet suggesting that they still have their own coherence.

Figure 23 is a graph of hashtags from the Libyan revolution. Similar to the Egyptian revolution, the roughly 360,000 tweets we gathered from February 20 to May 17 contained 4,433 distinct hashtags. As in Egypt, there were two major communities in this graph: one where the largest node was #libya and one where it was #gaddafi. Like Egypt, Libya has its own "other countries" community (Figure 24), which was the third-largest in that captaset. The #egypt and #tunisia nodes are about 6% of the size of the main nodes (#gaddafi and #libya). It has a reciprocal link to the events in Egypt.

Putting these two sets of tweets together in Figure 25 reproduces this structure, even if it does strain the galaxy metaphor somewhat. The "other countries" communities connect the two global events. In both galaxies, the major communities' homogeneity remains, as do the secondary communities ("Mubarak" and "Gaddafi"). Their connections also remain proportional. We can at least partially quantify the relationships between events, something that could be parsed over time, and thereby identify their evolving relationships based on social media discourses. The discourse connects these events, yet on the large scale of Big Data (or Big Capta) analysis, their differences remain clear.

The Louvain method is an iterative algorithm: it looks for communities among nodes, then looks for communities in those communities, and then looks for communities in those communities of communities, and so on. What we see in Figure 25 is the broadest scale of communities. The smaller communities (the "dictator" communities in each galaxy) maintain their identity if we stop a step or two earlier. Network theory helps us discover and discuss the polyvalent nature of events: events are comprised of groups of actors performing actions, which coalesce on many levels into larger groups of actions that we also call events. Each individual user maintains his or her identity, but they also form larger structures by participating in events, which themselves are parts of larger events. In essence, every actor and idea is a thread, and when many threads merge together, they become an event. These events can be linked yet differentiated based on their discourse.

(TP) The database of tweets thus represents a geographically and temporally organized archive of utterances which can be read, analyzed, and visualized in multiple ways:

First, in their singularity, the flash of the human voice, the singular utterance at a given moment in time and space.

Second, synchronically, as the voices, sentiments, or expressions of a single day, or perhaps in a single hour or minute.

Third, diachronically, as change over time, which can be considered by following a single person tweeting over many hours, days, or weeks, or structural trends and patterns emerging over time.

Fourth, as the aggregation and visualization of all the data in the database captured by the application. In every case, the archive is "less" than the event and functions primarily as a possible but always incomplete documentation of the past through the contingency of the present.

But there is something else that can be found in the "zoomed out" approach to analyzing the archive, namely the openness to the future and the identification of points of inflection or tipping points that might indicate something about future possibilities. As Charles Péguy writes: "Events have critical points just as temperature has critical points—points of fusion, congelation, boiling, condensation, coagulation, and crystallization. And even within the event there are states of surfeit which are precipitated, crystallized, and determined by the introduction of a fragment of the future event" (quoted in Deleuze, 53). Such points of inflection or states of surfeit cannot "predict" the future or manifest the future in any concrete or particular way; they merely introduce, in Péguy's elegant words, a fragment of the future event.

Our Egypt dataset is not quite large enough, not quite accurate enough (given that we cannot verify geographic locations provided by users), and not quite complete enough to definitively detect such inflection or tipping points. Nevertheless, we wanted to know: might there be some discernible statistical differences in the hours leading up to February 11 at 6PM, the hour Mubarak delivers his resignation speech? Might more (or less) people be tweeting on these days than in all previous days? Might there be a change in frequency in which people tweeted? Might we be able to extract topics from the tweets and see if certain keywords or topics (such as violence or fear) were becoming more intense over the period of the revolution? In other words, could we detect any statistical differences in the social media data that might give an indication that the revolution was reaching some kind of inflection point? The ramifications were genuinely exciting but also deeply sobering: massive social media datasets could be aggregated and studied to track the course of events that were in the process of unfolding and whose future remained unknown, but who is to say such statistical processing won't be used to enforce a particular future or eliminate the openness of the future to come?

(DS) While no one could predict (from the data) that Mubarak would give his resignation speech at that hour, on that day, we do see some changes on February 10 and 11 in comparison with all other days of the revolution, something that might have prompted policy analysts to start paying attention at this moment. Interest in the Egyptian revolution among Twitter users swelled during those two days, partially due to a number of events that turned the tide in the protesters' favor. Figure 26 shows that more tweets were sent on February 10 and 11 than on any other day we captured during the revolution. February 11 was the day the most tweets were sent, with the highest number and greatest frequency within the hours leading up to and following Mubarak's resignation.

On Twitter, these intensifications frequently came through retweets of particular—messages that form a significant portion of our archive (between 40–50%) on these two days. One significant tipping point during those two days was the Egyptian Army's decision to join the protesters. Several officers defected on February 10; from 10 AM to 12 PM, the most-retweeted tweet was "@arwasm: 3 army officers give up their ranks and join the protesters. One says 'My oath was to #Egypt, not to a person.' #jan25 #tahrir." Retweets of this tweet made up about 10% of the tweets sent during those two hours. Midway through the day, the Army stopped opposing the protesters ("RT @bencnn: From

(26) Capta-set of tweets by day.

Communiques 1 and 2, it's clear the Army is confused. This is getting muddled beyond words. #Egypt #Tahrir #Jan25"; "RT @NadiaE: Army not threatening thousands gathered at presidential palace. If this continues they have my respect for that #jan25 #egypt"; "RT @occupiedcairo: Army not doing anything to prevent ppl accessing TV building. All peaceful so far #jan25"). Midafternoon, the Army signaled sympathy with the protest movement: "RT @justimage: Army general on other side crying and shaking hands with protesters. #jan25"; "RT @justimage: All soldiers on other side of fencing at tv bldg look sympathetic to protesters. Very emotional scene. #jan25." The military's shifting loyalty could not help but be felt on Twitter. Well into early February 10, tweets blamed the Army for human rights violations, but as the Army made its new position known over that day, tweets expressed support for the Army. The volume of messages about this shift reflects how much of a tipping point this event signaled.

While each tweet or retweet is one individual user's speech act, it forms part of a community discourse, a complex network of speech acts. The large event, the

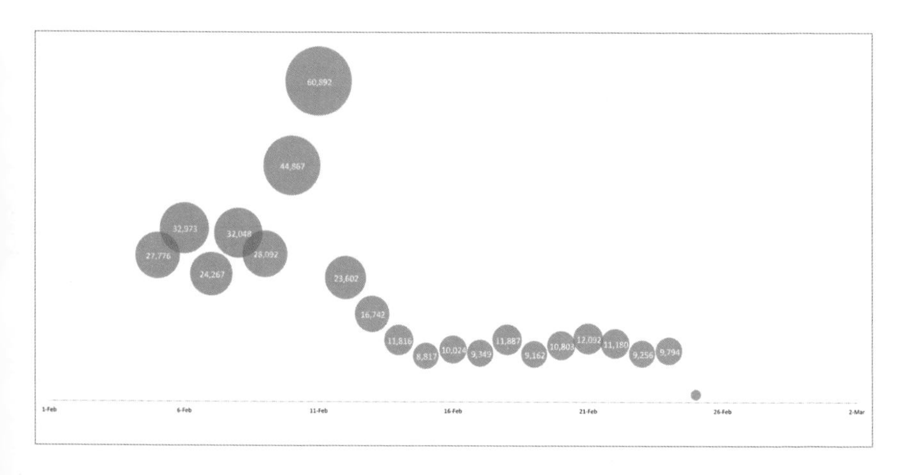

Egyptian revolution, consists of many smaller event threads and many individual experiences condensed through tweets. With visualization software to extract a global view of the network over time, we begin to detect event rhythms and even inflection points. If we view the Egyptian revolution as a network of people, actions, and ideas, we can simultaneously interrogate the revolution at a macro-scale and the many individuals' experiences as micro-threads that coalesce into and recursively affect the course of events, thereby giving rise to the possibility of a global public sphere of ever thicker (and ever more fragile) networks of participants, witnesses, and listeners.

Japan: Real-Time, Real-Space Disaster Relief from 5,000+ Miles Away

(YK) "Something terrible has just happened in Japan." Such was the message from my mother-in-law that I awoke to on the morning of March 11, 2011. Having friends and family in Japan, but living in Los Angeles for the past 15 years, I dread to hear such words—like many of us who live in countries foreign to our own. Turning on the television and browsing the web soon revealed that one of the worst catastrophes in our generation had just occurred in my home country, now more than 5,000 miles away. As the day unfolded, so did the horrors of entire communities being washed away by the unfathomably powerful waves. Calling back home had mixed results: while those far from the immediate impact zone of Sendai were able to respond, those living in Tokyo were not. It was to be another excruciating 24 hours before I was finally able to confirm the safety of all my friends and family in Japan. But it took less than those 24 hours for me to hastily submit my resume to a technology volunteer agency called GIS Corps and be recruited and assigned as the GIS Lead for CrisisCommons, an organization that coordinates volunteers in crisis situations. CrisisCommons had just started their Japan mission, and I joined hundreds of technology volunteers from around the world, each and every one of us working selflessly toward the relief efforts. For the next 10 days, we—an international collection of faceless, motivated, and talented technologists—poured our hearts into gathering data, conducting spatial analyses, and contributing to the relief efforts from afar.[85]

The Role of Social Media

@Chaa_aaki
3/12/2011 16:37:41
叔父の家が津波で浸水し、二階に二名取り残されてます！助けて下さい 石巻市築山3-2-26 #j_j_helpme ！！

> My uncle's house is underwater because of the tsunami and there are 2 people stranded on the second floor! Please save them. 3-2-26 Tsukiyama, Ishinomaki City #j_j_helpme

(YK) A full day after the earthquake struck, @Chaa_aaki's plea for help was seen on Twitter. It came from a woman trying to save her uncle, trapped on the second floor of his house that was located in a flood zone caused by the tsunami in Ishinomaki City. She added the hashtag #j_j_helpme, which was designated to be used by people seeking help in the aftermath of the earthquake. Her plea for help was retweeted, over and over again. She even left an address that would allow relief workers to find her uncle.

http://giscorps.org
http://crisis commons.org

Looking at the location on a map, sure enough, we find out that her uncle's house was just a few hundred meters from the shore, one among thousands of homes inundated by the tsunami.

While I do not know if this particular tweet actually mobilized relief agencies to save her uncle, Twitter—as a platform—was an important conduit for real-time, real-space information dissemination, used by both the general public and disaster relief agencies. Following the earthquake in Haiti in 2010, tweets became part of the data harvested by Ushahidi, a crowd-sourced visualization and mapping platform used to collect and coordinate information using multiple channels, including SMS, email, Twitter, and the web. This was taken to a new level in Japan, as the social fabric of the nation quickly revolved around the usage of Twitter as the primary mode of communication for requesting medical aid, seeking information about missing people, sending encouragement, and reporting damage and transportation infrastructure statuses. In fact, more than 80% of the 20,000 incident reports collected by Sinsai.info, the Ushahidi instance for the Japan disaster, came from Twitter. Over the next 30 days, more than 700 million tweets

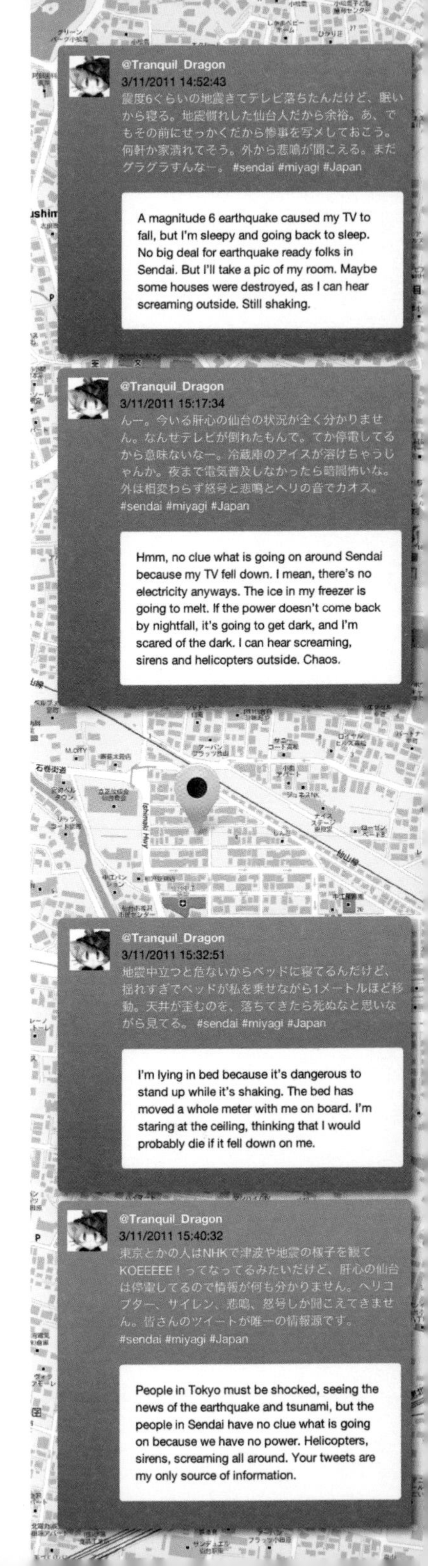

were sent. It is no longer a question of whether or not social media can be used in disaster situations, but rather a question of how they can influence actions and simultaneously document what is happening from a nearly infinite multiplicity of perspectives and locales.

Two days after her initial tweet, a follow up message from @Chaa_aaki appeared on Twitter:

@Chaa_aaki
3/14/2011 14:57:53
叔父、救助されたみたいです！本当に本当にありがとうございました！まだ救助されてない人が早く助かるようにお祈りします!!! @Chaa_aaki: 叔父の家が津波で浸水し、二階に二名取り残されてます！助けて下さい 石巻市築山3-2-26 #j_j_helpme ！！"

My uncle has been rescued! Thank you so very, very much! I pray that those that are still waiting for help will be saved!!

The Retweet

(YK) Part of the intrigue and power of the social web lies in its ability to transmit data through a multitude of networks that grow exponentially the more "popular" the information is. In Twitter, this is accomplished by means of retweeting (or RT) capabilities: the simple operation of sharing a tweet with others in your network and, subsequently, enabling people within your network to retweet it

@harukalunch
3/14/2011 1:27:36
RT @NamicoAoto: 父が明日、福島原発の応援に派遣されます。半年後定年を迎える父が自ら志願したと聞き、涙が出そうになりました。「今の対応次第で原発の未来が変わる。使命感を持っていく。」家では頼りなく感じる父ですが、私は今日程誇りに思ったことはありません。無事の帰宅を祈ります。#jishin

RT @NamicoAoto: My eyes filled up with tears when I heard that my father volunteered to go to the Fukushima Nuclear Plant, even though he will be retiring in just half a year. He said that "the future of this nuclear crisis depends on what we do now, and I must go." At home, he is not always the most reliable father...but today, I have never felt as proud of him. I pray for his safe return. #jishin

to their own networks until a tweet reaches a massive audience, sometimes in a matter of hours. In the case of the tweets related to the earthquake, retweets were used effectively to broadcast infrastructure damage, post missing person notices, and even announce relevant hashtags.

@NamicoAoto's tweet was retweeted more than 17,000 times. While most tweets were informational in nature, the most "popular" tweet was about courage and sacrifice. This ideal of sacrifice resonated deeply during trying times, even prompting Prime Minister Naoto Kan to proclaim to the volunteers, "You are the only ones who can resolve a crisis. Retreat is unthinkable." The international media called them the "Fukushima 50," heroes who volunteered to stay behind at the crippled Fukushima Dai-ichi nuclear plant in order to prevent a meltdown in Japan.[86] Locally, however, the complexity that surrounded this "man-made" disaster caused many to refrain from celebrating and lauding these men. Heroes or not, it would not be until October of 2012, a full 18 months later, before the next Prime Minister Yoshihiko Noda finally honored and thanked them for "saving Japan."[87]

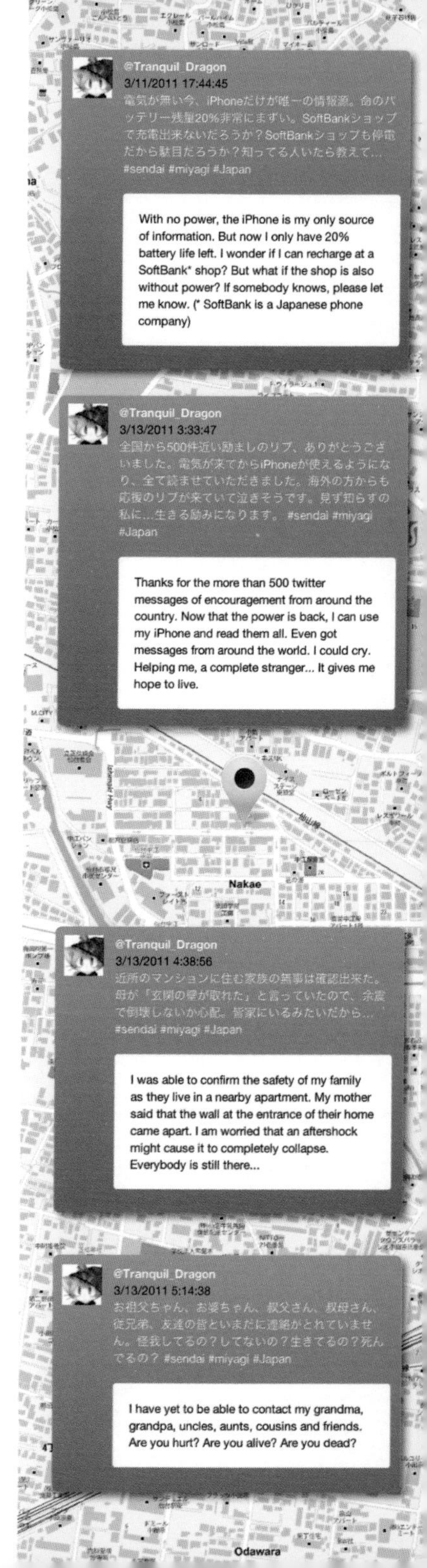

A Zoomed Out View of Twitter Data From Japan

(YK) Kawano and Shepard developed several web-based visualizations that seek to expose spatial, temporal, and lexical patterns that are otherwise dormant in the vast amounts of data captured in millions of tweets. These visualizations reveal real-time patterns through synoptic views of the data that are not perceivable in the linear and ephemeral feeds coming from the Twitter website interfaces. At the same time, the team used the perspectives and patterns of the data in aggregate to "zoom in" to particular individuals and follow individual stories unfolding.

UCLA's Japan Twitter archive was collected over a 35 day period, from March 9 to April 13, 2011, via a "cron job"[88] that queried Twitter's search API every 3 minutes, collecting tweets based on a location parameter (Japan) and at least one relevant hashtag (#earthquake, #sendai, #jishin, #tsunami, #eqjp, #pray4japan, #japan, #j_j_helpme, #hinan, #anpi, #daijyoubu, #311care). The tweets were subsequently saved in a database on a server at UCLA. While the archive has close to 700,000 records, it is a small—but focused—portion of the supposed 700 *million* total tweets sent during the same time period. Here's a look at the raw numbers:

689,054	total number of tweets collected
236,520	distinct users, with 2.9 average number of tweets per user during this 35-day period
573,841	(83%) retweets (with the word "RT" in the text)
199,741	(29%) distinct tweets (some of which are also retweets or duplicates with slight variations)

@aroe_s, who has a Twitter space dedicated to information regarding earthquakes in Japan, tweeted more than 2,500 times during this time period (by far the most prolific user). The highest number of tweets per hour comes about a month after the earthquake: April 7 at 11:32pm. This is assuredly due to the occurrence of the second largest aftershock that shook Japan at magnitude 7.1 (there was actually a 7.9 earthquake that followed 30 minutes after the main 9.0 earthquake on March 11).

Sentiment Analysis

(YK) One aspect of social media is that its ephemeral nature can provide a snapshot of a moment's mood, reflected—in aggregate—by the content of what people are tweeting about in real time. The rise and fall of different emotions are a function of the many events that transpired in the aftermath: the evolving story around the nuclear meltdown and radiation contamination, the timing of the many devastating aftershocks, the increasing death toll, and the futility of finding survivors all contributed to the nation's psyche. In order to analyze the emotional and psychological state of the nation in the

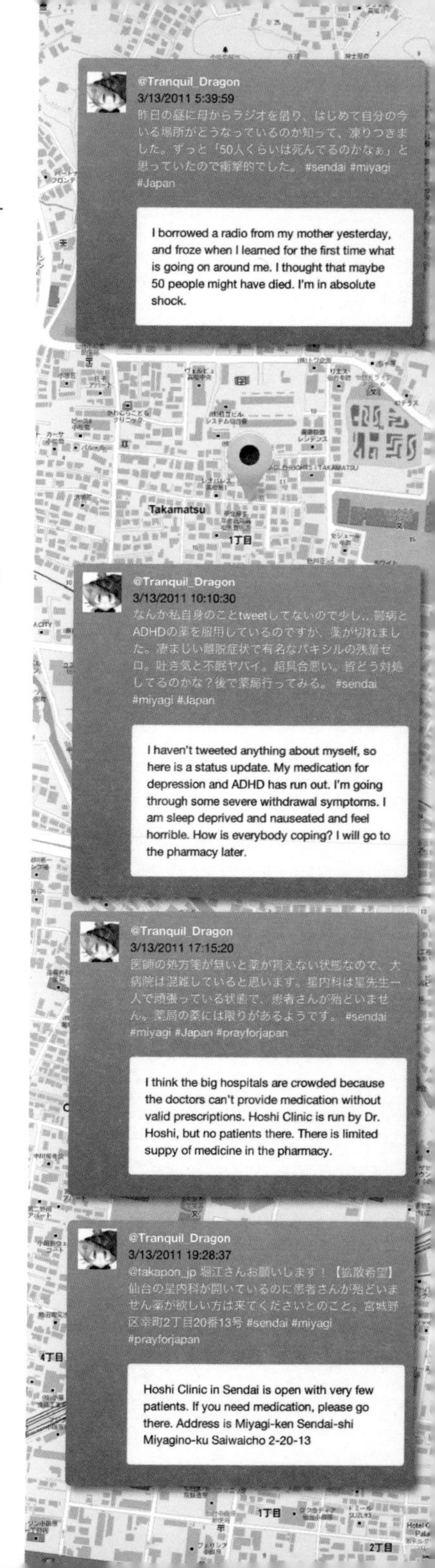

days after the disaster, the archive was divided into 35 separate "day" text files. To measure day-by-day fluctuations of emotions, a methodology that takes more than 2,000 of the most "emotional" words in the Japanese language (as defined by the "Emotion Expression Dictionary"[89]) was matched against each of the "day" files to gauge the emotional content.[90] The dictionary classifies these words into ten meta-categories: weariness, happiness, restlessness, sadness, fondness, fear, anger, peacefulness, surprise, and shame. By aggregating the match rate per day, Twitter's data is transposed into a flow of emotions along a temporal pathway.[91]

So what was the country "feeling" after the earthquake? Was it engulfed in sorrow? Anger? Fear? What effect did the hundreds of aftershocks have on the populace? Creating a universe represented by two object types—days (in black) and words (colored by emotion categories)—one can see the gravitational pull for the more "central" days and words. (Fig. 27) Unlike the multiple centralities that exist in the previous Egyptian universe, this one has a singular core, dominated by objects that orbit around a single, focused event. Words are sized by how often they were spoken, and their correlation to the different days is represented by lines that are also sized according to the usage per day. Communities are created by clusters of words, which are mentioned in proximity to each other. The two largest clusters are represented by the words 涙 (tears) and 誇り (proud), which have formed a giant galaxy around March 14 and 15. This reflects the content from @NamicoAoto's single tweet (that was consequently retweeted more than 17,000 times) and embodies the nation's anxious sentiments, days after the earthquake, expressing pride and sadness in the same context.

The next largest cluster is formed by the word 心配 (worry), a sentiment that is classified as "fearful," and is marked by its presence following the two largest earthquakes, on March 11 and April 7. The final cluster community revolves around March 13, two days after the earthquake, showing strong connections between the words 悲しみ(sadness), 恐怖(scared), 不安(anxious), 愛情 (love), and 傷(pain).

But what about replaying these emotions over time? The charts in figures 28/29 reflect the 10 emotion categories over the 35 day period. Looking at the top chart, many of the sentiments are dormant during the first couple of days, but collectively explode shortly after. Perhaps this is because the initial "shock" of the event prompted content that was largely informational in nature. "Happy," "sad," and "restless" sentiments each peak three days later, on March 14 ("happy" peaks here because the word "pride," from @NamicoAoto's tweet, is classified in it). But "fear" does not reach its high point until April 8, the day after the second largest aftershock. The country experienced many earthquakes over the course of the month, averaging more than ten earthquakes a day. However, the number of earthquakes had tapered

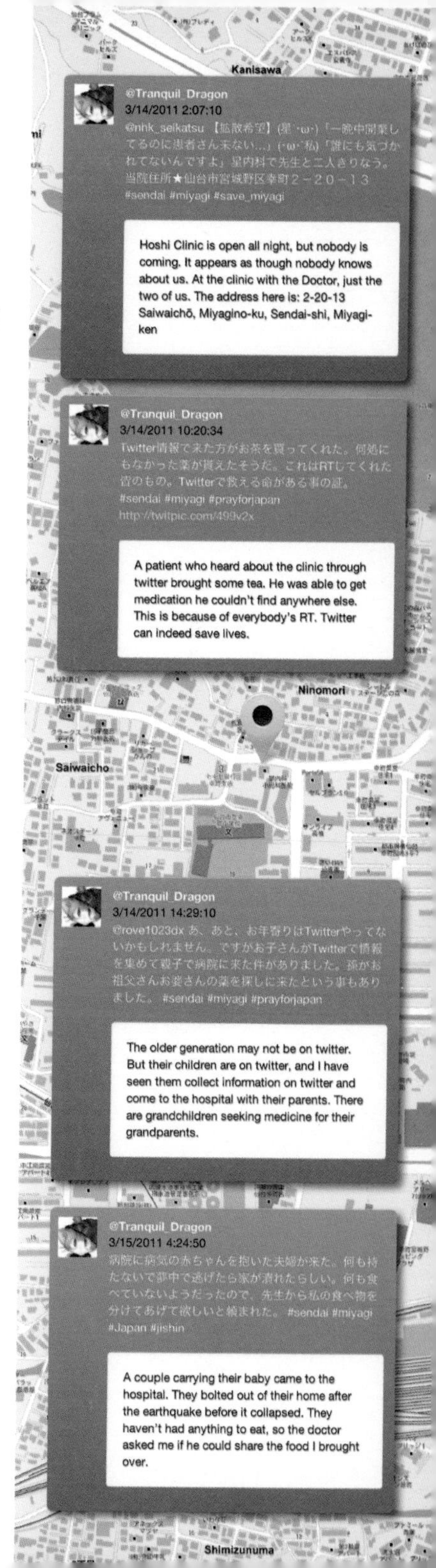

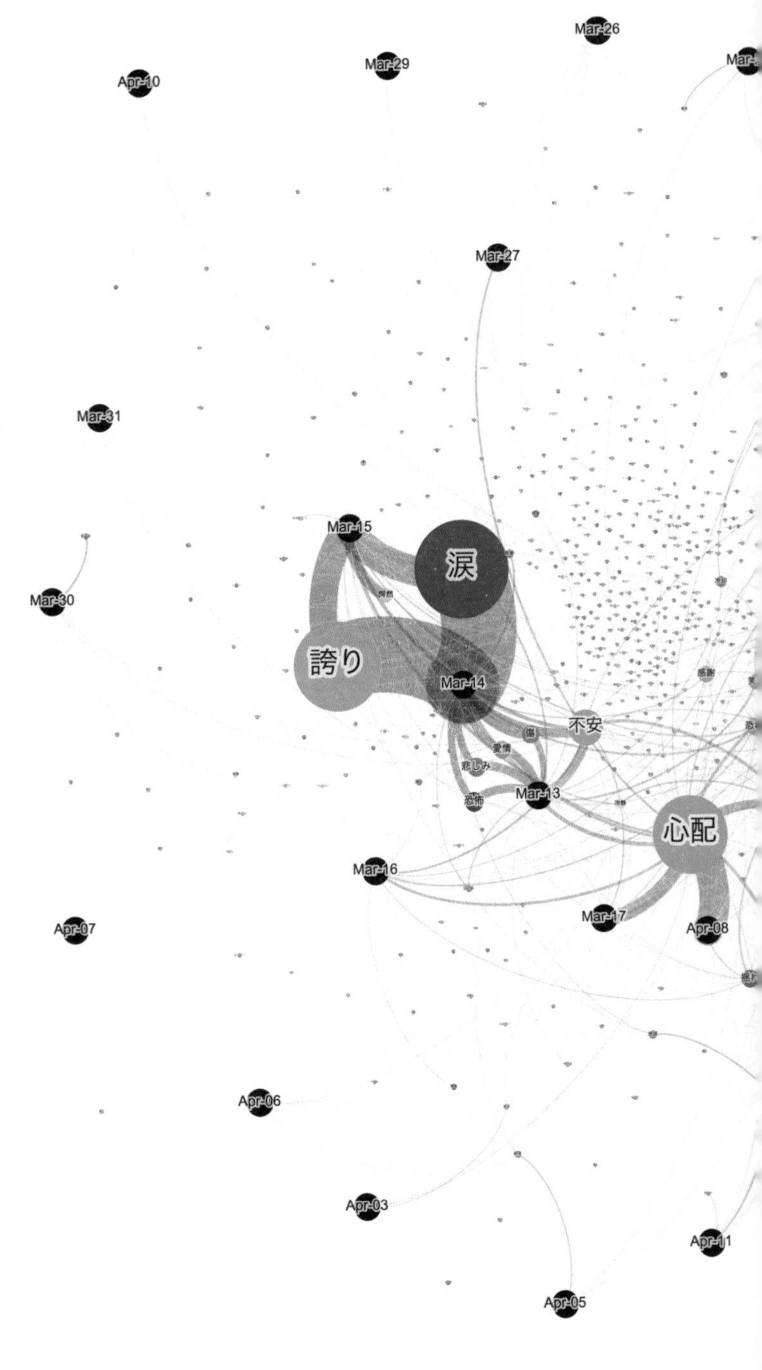

(27) Network diagram showing emotion keywords that were prevalent after the earthquake.

Mar-21
Mar-28
Mar-20
Mar-18
Mar-22
Apr-01
Mar-19
Apr-04
Apr-09
Apr-02
Mar-10
Mar-09

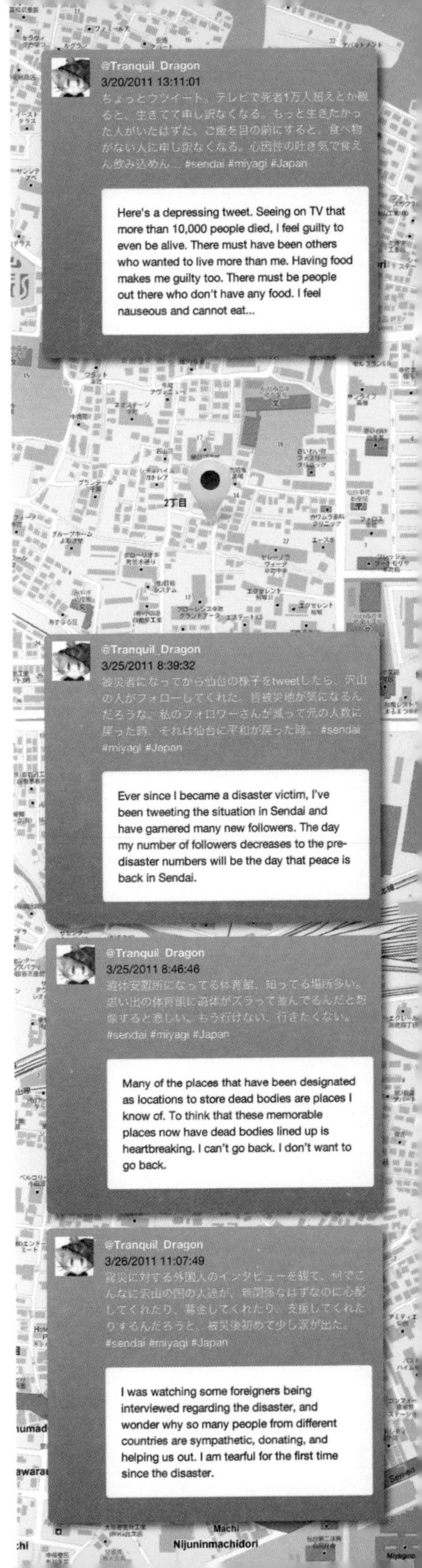

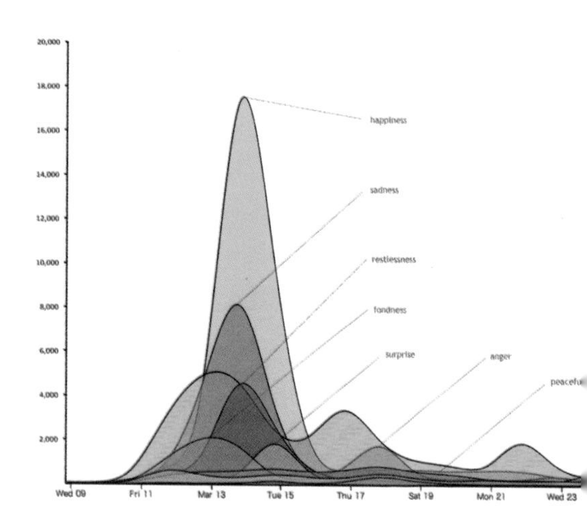

off considerably over time, perhaps explaining why the April 7 earthquake rekindled and renewed much of the fear in a nation that may have just started to begin its healing process.

While this chart effectively communicates the sheer quantity of emotional content, it does not reveal the distribution of the 10 emotions on a day-by-day basis and gives a false impression that the nation had no emotions between March 23 and April 7. To be sure, there was—measured in absolute numbers—significantly less emotional content in the Twitter archive during this period than in the days immediately

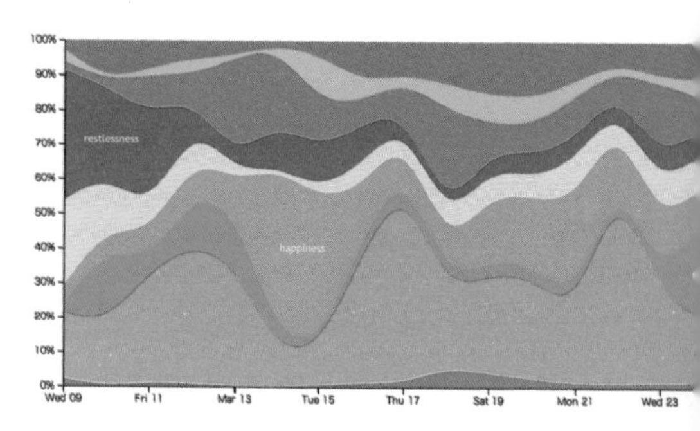

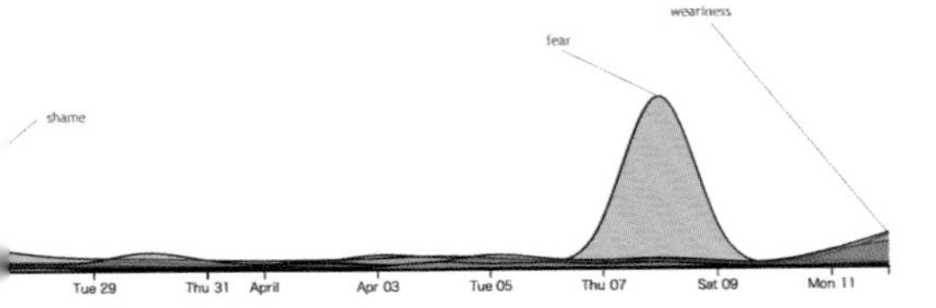

(28) Emotions charted over time.

following the earthquake and tsunami. As such, the chart below was generated to better expose the significant differences between each emotion group, regardless of the total tweet count, as values were normalized and scaled to 100%. One can see that "fear" is a dominant sentiment throughout and "weariness" lingers and escalates with time. Both are expressions that bespeak an exhausted population dealing with the uncertainty of the on-going aftershocks and, at the time, the unbounded and uncontrolled nuclear catastrophe.

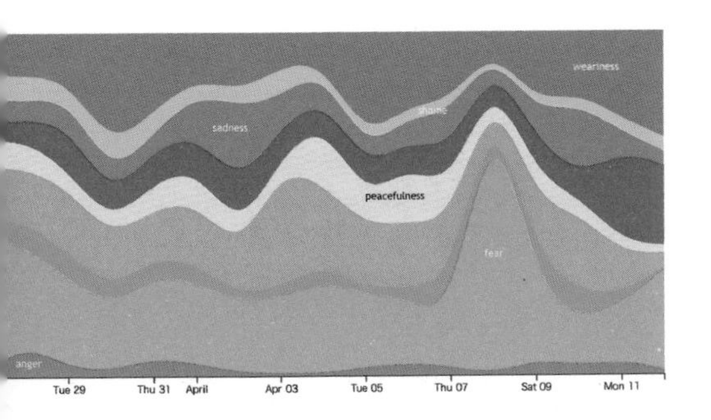

(29) Emotions as relative percentages over time.

A journey through Tohoku, Japan

"That building is on its side," I thought to myself in disbelief as I visited the city of Onagawa in Tohoku, Japan. Living abroad gave me the opportunity to analyze the "big" picture, taking a macro-level approach and applying computational analysis on a large, even abstracted and high-altitude scale. But now I was on the ground, walking among the ruins. While it is not possible to turn back the clock and experience the disaster as an eyewitness, it is possible to mobilize and immerse oneself into the physical spaces that still define the specter of the event.

I first met Yugo Shobugawa at Harvard University in May 2011. He came to watch our team's presentation on "Social Media and GIS: Mapping Revolutions, History, and Catastrophe" where I spoke about the role that social media played in the days following the earthquake. Yugo is a physician, an expert in clinical epidemiology, and currently teaches at Niigata University. He is also a GIS practitioner, fluent in using spatial analysis to provide insights on how physical space is directly correlated with the outbreak, diffusion, and control of various epidemics. We bonded immediately over mutual excitements like temporal mapping and spatial regressions. But I think the real connection happened

when we both acknowledged our feelings of despair and helplessness of being Japanese, and yet not being in Japan during the earthquake. For many Japanese nationals living abroad, experiencing the earthquake from a distance was excruciating. It is this sense of despair that drove many of us to pour our hearts and souls into the recovery effort remotely.

When I finally had the opportunity to return to Japan, on two separate occasions—once in December of 2011 and again in July of 2012—I decided that I had to see the effects of the disaster with my own eyes. I had seen a similar catastrophe in the past, having visited the town of Banda Aceh, Indonesia, shortly after the Indian Ocean Tsunami in 2004. This trip left a huge imprint, witnessing an entire community in ruins, standing on ground zero where more than 100,000 people died, rendered speechless and heartbroken. With much trepidation, I understood that going to Sendai would bring similar sentiments in some ways, but profoundly different ones in others, given my deep cultural and spiritual bond with the country. I called Yugo, and together we took several trips to the disaster-struck areas.

Yuriage Elementary School

A missing baby poster is about the saddest thing that can greet you, but this was exactly what I saw affixed to the entrance to the Yuriage Elementary School gymnasium. The piece of paper, attached with five pieces of tape, already had signs of wear and tear, as my visit was a full 9 months after the earthquake. The poster looked something like what you see below. The photo on the poster reminded me of my own baby boy who had just turned one. The boy's cheeks were flush, and he had a full head of straight black hair. His face was round, and he smiled with his mouth open, looking straight at the camera, looking straight at me. I stood there silently, transfixed by the image of the smiling boy sporting the popular shimajiro tiger bib. Many months later, I found out that the father had lost four members of his immediate family and was still searching for his baby boy.

Looking for
8-month-old baby
Height: about 70cm
Weight: about 9kg
Bottom set of teeth just
coming out

Beyond the door, I came to the "Fuji Film Salvation Photo Project," a volunteer effort to collect and recover "memories" from the disaster zone, and to carefully organize and exhibit them in hopes that family and friends could retrieve them. The room was filled with box after box and crate after crate of photographs, diaries, scrapbooks, and personal effects. It was the physical archive, on paper and in photographs. Items were tagged and labeled by location, date, names, and whatever else could be extracted during the recovery process. The memories of the past from so many distant locations had been gathered together in this one place. Sometimes, it was all that remained of families, evidence that they had existed, that they had lived. These photographs—unclaimed—were all that was left.

Christmas at Okawa Elementary School

When we visited Okawa Elementary School on December 20, 2011, it was a dreary, overcast day, with snowflakes falling from the sky in a slow swirling motion. Of the 108 children in attendance, 74 were lost due to the tsunami that swept through the school and much of Ishinomaki City. The day of the earthquake supposedly had similar weather, with temperatures hovering around freezing levels. As I took out my camera, I could hardly keep my hands on the shutter, as the cold wind cut deep into my exposed fingers. Empty spaces that showed remnants of the disaster surrounded the entire school: squashed cars swept away and left abandoned in the middle of rice paddies, mounds and mounds of stacked debris, an occasional empty house that miraculously survived the waves, bulldozers continuing the cleanup effort, smoke rising from place to place signaling the burning of collected trash. The school itself was wedged between the bank of a river and a small hill on its backside.

According to our local guide, Satoshi Mimura, a medical officer in Japan's Ground Self Defense Force and current PhD student at Tohoku University, the earthquake manual for the school dictated that the children evacuate the building and congregate in the open area outside the building. This is the area that is sandwiched by the river and the hill. The school teachers kept the children in this playground, instructing them to await their parents who were on their way to pick them up. Some parents did show up, taking their children with them. They were among the few who survived. Shortly after, the first warnings of the tsunami came. After much debate, the instructors chose to march the children toward higher ground, but made a tragic error in judgment by deciding to head toward a nearby bridge, where they were met head on by the dark, powerful waves.

Four of the children, along with one of the teachers ran toward the hills in the opposite direction, reaching higher ground, clinging onto trees and most likely watching helplessly as their less fortunate classmates were swallowed up by the waves. Those who survived had to endure freezing temperatures overnight before finally being rescued from the hill the following day.

What is left standing in Okawa Elementary School is the shell of the main campus, a brick building whose main circular structure has remained intact. In front of the structure is a memorial, adorned today with fresh flowers, toys, stuffed animals, fresh fruit, drinks, and funerary plaques. In front of the memorial are two poster boards. The first one reads:

Please look after us
Fathers
Mothers
Little by little, little by little,
We will move forward.

This is when I noticed the solar panels located next to the memorial. These panels were powering the illumination for the Christmas tree that stood in the exposed wall behind it. A translation of the second poster appears below:

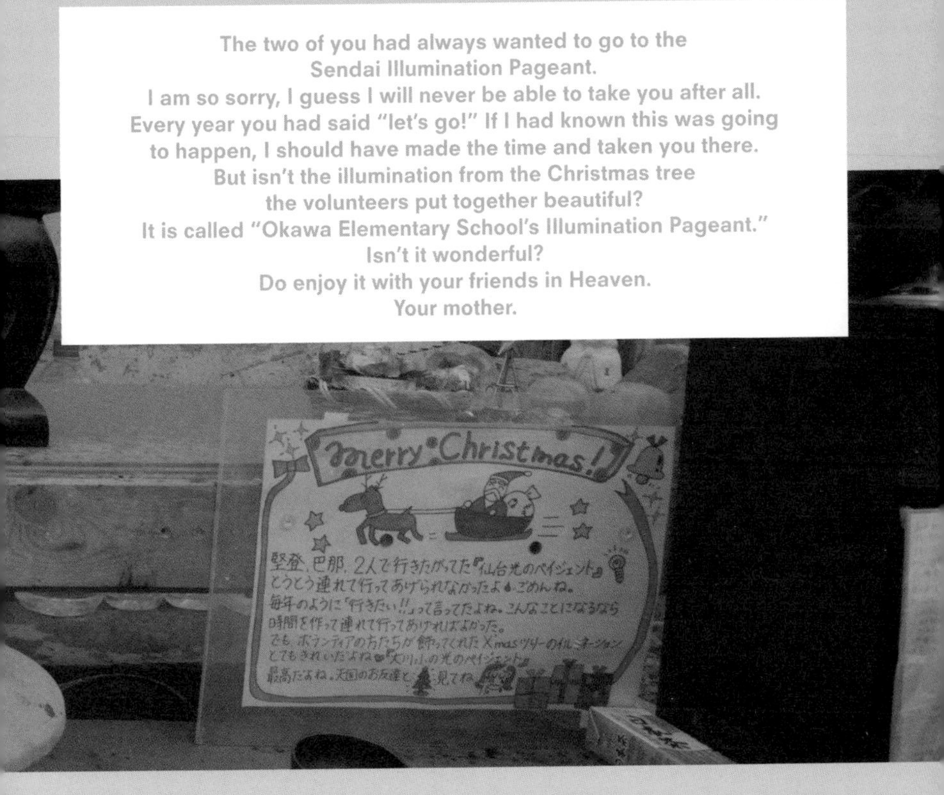

The two of you had always wanted to go to the
Sendai Illumination Pageant.
I am so sorry, I guess I will never be able to take you after all.
Every year you had said "let's go!" If I had known this was going
to happen, I should have made the time and taken you there.
But isn't the illumination from the Christmas tree
the volunteers put together beautiful?
It is called "Okawa Elementary School's Illumination Pageant."
Isn't it wonderful?
Do enjoy it with your friends in Heaven.
Your mother.

南相馬市
Odaka
浪江町
葛尾村
双葉町
富岡町
楢葉町
広野町
川内村
村市
399
114
459
288
6

Inside the nuclear evacuation zone

My second visit to the Tohoku region came in the summer of 2012. This time, I would be joining Yugo and his colleagues to venture inside the 20km nuclear evacuation zone, a place where more than 170,000 people had been forced to evacuate following the declaration of a nuclear emergency by the government at 7:03 PM on March 11.

As the days counted down toward my visit to the barricaded areas, I began to wonder: what would it be like to physically enter a place where only a tiny handful of people on this planet have been allowed to visit since 3/11/11? The public outcry and the media frenzy that followed the ill-fated Fukushima Nuclear Plant has put a spotlight on this location, serving as a reminder to the world of what a deadly confluence of natural disasters and human incompetence can do. According to a recent 600 page report[92] conducted by Kiyoshi Kurokawa, a Tokyo University Professor, on the causes of the nuclear catastrophe, he concludes:

> *What must be admitted—very painfully—is that this was a disaster "Made in Japan"*

July 17, 2012

My day began early in the morning as I hopped on the 7am "Max" bullet train from Tokyo station to Niigata. Upon my arrival, I took a cab to Niigata University Hospital where Yugo works as a medical researcher.

I got off at the in-patient lobby of the hospital, a place I later learned had received hundreds of patients from Fukushima shortly after 3/11, some of whom died shortly after they arrived. Following the earthquake, many hospitals were overburdened and doctors were short in demand, and so, even this hospital, located 4 hours (170 miles) from the Fukushima Nuclear Plant, received an influx of patients.

I soon met up with Yugo and was introduced to his many colleagues. Yugo leads a team of researchers, working in International and Public Health. He then took me to another wing on campus called the Radioisotope Center, where he introduced me to the Bishamon team.

I am a member of the Bishamon team, through a partnership between Niigata University and UCLA. Our GIS and Visualization team (that consists of David Shepard and myself) at UCLA's Institute for Digital Research and Education has provided a web visualization strategy for the information being collected by the Niigata team.

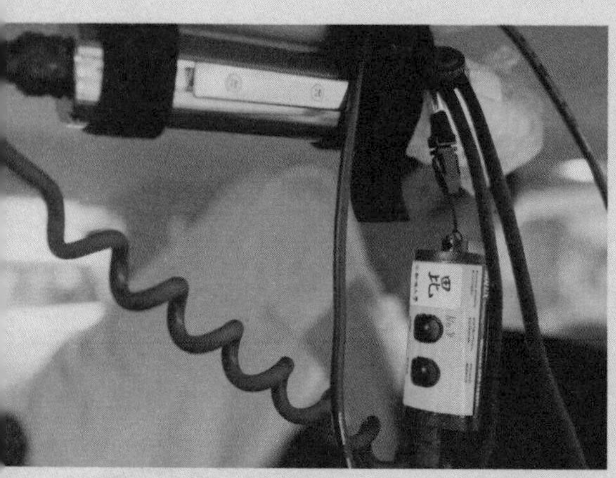

http://bishamon.org

While the Bishamon Niigata team gathers data on the ground, the Bishamon UCLA team processes and loads the data into a custom-built web application that provides the information to the public.

Bishamon is comprised of a group of sixteen academic volunteers, each an expert in their respective fields: radiology, nuclear physics, nursing, virology, community medicine, pharmacology, biochemistry, public health, and information technology. Most of us had never met each other before the disaster. On this day, four of us were to embark on a road trip to visit several cities that fall within the evacuation zone. Yugo specializes in epidemic diseases. Jun Goto, who never stops smiling, is a nuclear physicist who I soon realized was the genius behind the many devices concocted for the project. Makoto Naito is a doctor specializing in pathology and the current director of the Radioisotope Center, and I should mention, the charismatic leader of this band of volunteer researchers. Makoto is the heart and soul of the team, not only because he serves as the project director, but more so because he is from Minamisoma, the largest city that falls within the evacuation zone, symbolizing the roots and wings that are at the core of much of the work conducted by the team.

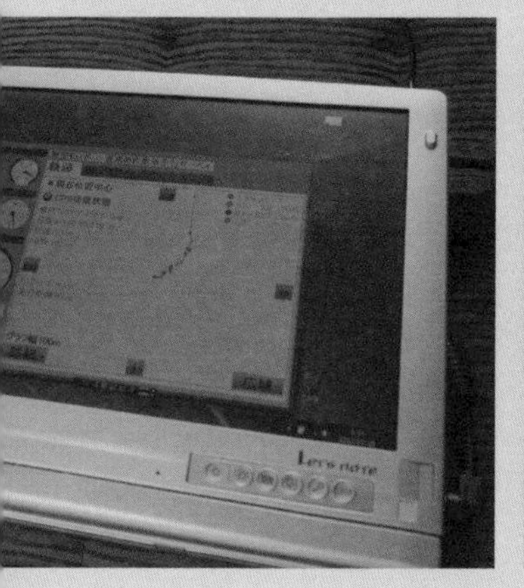

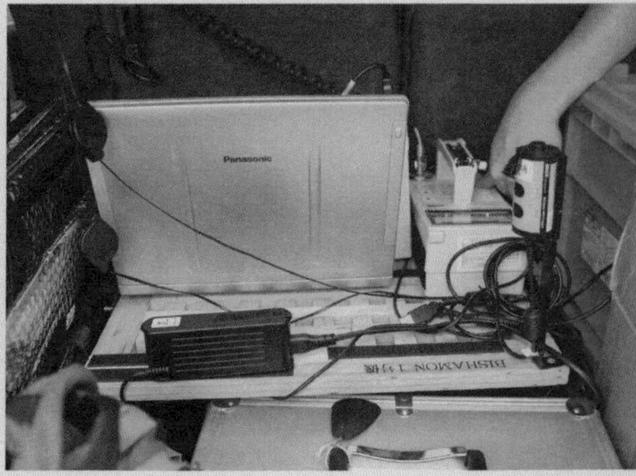

So what exactly is Bishamon? While it serves as the name and identity of the research team, it refers to the device that Jun has put together that measures airborne radiation levels through a mobile and vehicle mounted sensor. Simply put, it combines a dosimeter, which measures radiation, a GPS, for position and altitude, and a laptop, which houses a custom built application that compiles, stores, and visualizes the data streaming from the multiple devices. What makes Bishamon unique is the relentless pursuit of accuracy and consistency. The team believes in a single measuring protocol that can be replicated for years to come, if not decades into the future. This approach comes with the understanding that this is a long-term project, and that measurements need to be made, possibly over and beyond the life of an entire generation, to accurately understand the effects that the nuclear crisis has had upon the populace it affected.

The project itself has a simple mission: to empower local governments to measure remnants of radiation within their communities and to inform the residents and the general public of existing levels of radiation near and around their homes. While other projects may not have the capacity to measure over an extended period of time, Bishamon plans to be around for a while. The transfer of technology from research universities to local governments is at the core of the project's objectives. For local governments, having the capacity and the procedural know-how to measure radiation on their own is imperative to conduct targeted and prioritized decontamination efforts that can accelerate the process of healing and ultimately allow residents to come home. For the residents, this information can serve as a reality check and a factor for making informed decisions. While the demand for this information has accelerated, little (to date) has been done by the federal government to make the data available to the public in a timely fashion. In this regard, the Bishamon team intends to provide full transparency and access to all the data collected, both locally and to the international community. The "openness" of the data remains an important, long term goal of the team. Given the unprecedented nature of the disaster, it is imperative for the world to know what has and continues to transpire in Fukushima.

And so it was on the day of July 17 that the four of us took off from Niigata University heading toward the Fukushima Nuclear Zone. While this marked my first physical encounter with the Bishamon device and the first time I was part of this drive toward Fukushima, it was one of more than 25 such trips that the core members of the Niigata team had taken since the summer before, an almost biweekly excursion.

July 18, 2012

Spending the night in Iwaki, just south of the nuclear zone, we woke up early and had breakfast at our hotel. One thing that caught my attention was the rice. There were two options: Fukushima rice and Akita rice, perhaps, to give guests a choice between a possibly "contaminated" local Fukushima brand and another option that was surely "safe." It symbolized the existing sensitivity toward the nuclear situation.

After consuming the local Fukushima rice for breakfast, we took off to enter the Nuclear Zone. We put on our radiation protective suits, an act that condensed the gravity of the moment and also seemed to forge a sort of unwritten code of solidarity between the team and the people who had been evacuated from this region. We proceeded to enter from the south, heading north from Iwaki. The entrance was heavily barricaded, and I was surprised to see the large contingent of security officers present.

Once inside, it literally felt like we were in a bubble, a time capsule that had descended upon this entire land, filled with crumpled buildings, overgrown shrubbery, and debris that had been left untouched since March 11, 2011. While there was some traffic on the road (military, police, decontamination workers), there was not a single soul occupying any

of the residential or commercial structures around. This space was simply devoid of any real signs of life. In the eerie quiet, I was struck by how incredibly beautiful the landscape was inside. Unlike urban Japan, this region was representative of the nature that abounds in rural Japan, with gently winding roads, small bridges, fog clinging to the mountains, a myriad of rivers, an agricultural landscape dominated by rice fields, and a blend of traditional and modern Japanese architecture.

Beep, beep, beep. All along, Jun had the Bishamon device turned on inside our vehicle. The real-time monitoring system was measuring radiation levels, producing an audible beep every second that a new reading was recorded. We silently stared at the charts and maps being generated on the computer screen as we approached the nuclear power plant. 2 … 10 … 15 … 20 … micro-sieverts per hour, readings that were deemed way above any safety guidelines by any standards. We came to within 2km of the nuclear plant before veering off toward the ocean, headed toward Ukedo Elementary School. As we approached the ocean, we noticed cars, boats, and all sorts of debris strewn across the landscape, many as far as a few miles inland.

Ukedo Elementary School

Today, Namie City is a city that effectively no longer exists. All 21,000 residents were forced to evacuate shortly after the nuclear meltdown, and none has been able to return. Satoru Shirato, Hideyuki Matsumoto, and Michiko Takahashi are government employees for the City of Namie working in a make-shift office in Nihonmatsu City. They kindly accompanied and guided us through what was left of their city, which is almost entirely contained within the evacuation zone. Our first stop was at Ukedo Elementary School.

3:38pm. "That is odd," I thought, as I noticed that every

clock in every classroom had stopped at exactly 3:38pm. And then it hit me. The earthquake happened at 2:46 pm. However, I soon realized that the clocks did not power off because of the earthquake, but rather, by the tsunami waves that had pounded the facilities 52 minutes later. Every classroom we entered reflected this fact (which also demonstrates the Japanese accuracy in synchronizing their clocks). Each classroom also had blackboards that showed the un-erased date (3/11) and messages of encouragement left by the relief workers.

The school grounds are locat-

ed just a few hundred meters from the shore, and the facilities evidenced the staggering devastation left behind by both the earthquake and the tsunami. As I entered the gymnasium, I almost fell … and noticed that the floor had caved in, exposing the raw construction materials like spikes on the floor. The banner inside the gym read: "Congratulations, new graduates." According to Mr. Matsumoto it had been a festive mood on the morning of March 11, where rehearsals for graduation ceremonies were being held in the gym. Fortunately, there were no student casualties at Ukedo Elementary School, unlike the tragedy that befell Okawa Elementary School. According to him, upon hearing the tsunami warnings, instructors commanded the student body to evacuate to the nearby mountains, where a major road (Highway 6) lay on the other side. This was about a 2.5 km trek, and many of the students were then picked up by passing trucks, where they were transported to nearby evacuation shelters.

I acquired multiple perspectives during this tragedy that connect me to the many people, places, temporalities, and mediations of this singular event, first as a spectator from the outside working feverishly with those inside, and then as a witness from the inside who could leave and return to my outside. The event has many lives and afterlives through haunting absences, stranded memories, and empty landscapes. I find myself moving between the gaps: assuming a top-down view of collecting and gathering digital data, applying computational algorithms in an attempt to understand an objective "whole," and the intensely personal, subjective, and emotional attachment one acquires by "being there."

Back in Los Angeles, I continue to listen: analyzing, mapping, and visualizing the data from the beep, beep, beep, while my colleagues return, time and again, to the nuclear evacuation zone.

Yoh Kawano

We follow the story of a single eye-witness, Tranquil Dragon, through the streets of Sendai in the days, weeks, and months after the disaster.

What happens to the memories of the children from Okawa elementary school, of the baby gone missing, and the lives salvaged, however hastily and momentarily, in the crates of photographs and boxes of paper?

The event remains.

It is still possible to bear witness for the witnesses.

Endnotes

1 **Walter Benjamin**, *The Arcades Project*, trans. Howard Eiland and Kevin McLaughlin (Cambridge: Harvard University Press, 1999), 476 (translation slightly modified). The original German is: "Geschichte zerfällt in Bilder, nicht in Geschichten." *Das Passagen-Werk*, ed. Rolf Tiedemann (Frankfurt am Main: Suhrkamp, 1983), 596.

2 **Theodor Nelson**, "A File Structure for the Complex, the Changing, and the Indeterminate" (1965), in: *The New Media Reader*, eds. Noah Wardrip-Fruin and Nick Montfort (Cambridge: MIT Press, 2003), 133-45.

3 **Norman J.W. Thrower**, *Maps and Civilization* (Chicago: University of Chicago Press, 1972), 3.

4 **David J. Bodenhamer**, "The Potential of Spatial Humanities," in: *The Spatial Humanities: GIS and the Future of Humanities Scholarship*, eds., David J. Bodenhamer, John Corrigan, and Trevor M. Harris (Bloomington: Indiana University Press, 2010), 14-30. In practice, there is much convergence between the notion of "thick mapping" that we are describing here and Bodenhamer's "deep maps" that embody humanistic approaches to spatio-temporal data marked by ambiguity, uncertainty, contingency, and incompleteness. See his forthcoming edited volume: *Deep Maps and Spatial Narratives* (Bloomington: Indiana University Press, 2014).

5 **Fredric Jameson**, *Postmodernism, or, the Cultural Logic of Late Capitalism* (Durham: Duke University Press, 1992).

6 **Clifford Geertz**, *The Interpretation of Cultures* (New York: Basic Books, 1973).

7 http://www.artcom.de/en/projects/project/detail/the-invisible-shape-of-things-past/.

8 **Walter Benjamin**, "The Work of Art in the Age of Mechanical Reproduction," in: *Illuminations*, trans. Harry Zohn (New York: Schocken Books 1968), 236.

9 **Michael Jones**, "The Future of Local Search—Google's Strategic Vision," Conference Talk at Where 2.0 (May 29, 2007).

10 **Cathy Davidson**, "Humanities 2.0: Promise, Perils, Predictions," in: *PMLA* 123.3 (2008): 707-17; Cathy Davidson and David Theo Goldberg, *The Future of Learning Institutions* (Cambridge: MIT Press, 2009).

11 For Twitter archives from Egypt, Libya, and Japan, see, respectively: http://egypt.hypercities.com, http://libya.hypercities.com, and http://sendai.hypercities.com.

12 **Theodore H. Nelson**, *Literary Machines: the report on, and of, Project Xanadu concerning word processing, electronic publishing, hypertext, thinkertoys, tomorrow's intellectual revolution, and certain other topics including knowledge, education, and freedom* (San Antonio, TX: T.H. Nelson, 1987), 0/13.

13 http://salt.unc.edu/T-RACES/.

14 http://acl.arts.usyd.edu.au/harlem/.

15 **Edward Said**, *Culture and Imperialism* (New York: Vintage Books, 1994), 58.

16 Restoring the term "culture" to its etymological roots of cultivating, tilling, and inhabiting, Casey argues that "we must, finally, *put culture back in place.*" "How to Get from Space to Place in a Fairly Short Stretch of Time," in: *Senses of Place*, eds. Steven Feld and Keith Basso (Santa Fe, New Mexico: School of American Research Press, 1996), 34.

17 **Walter Benjamin**, "One Way Street," in: *Walter Benjamin: Selected Writings,*

Volume 1: 1913–1926, eds. Marcus Bullock and Michael Jennings, trans. Edmund Jephcott (Cambridge: Harvard University Press, 1996), 444–48.

18 In his seminal book, *The Image of the City* (Cambridge: MIT Press, 1970), **Kevin Lynch** articulated the ways in which mental geographies (personal associations, memories, desires) meld with and even structure the physical geographies of the city.

19 On the development of planimetric maps and the evolution of the bird's eye perspective, see **J.B. Harley** and **David Woodward**, eds., *The History of Cartography*, six vols. (Chicago: University of Chicago Press, 1987-92).

20 **David Wood** argues that the history of cartography is inseparable from the history of the nation-state in *The Power of Maps* (New York: Guilford Press, 1992). For an overview of the history of "cartographic reason," see **John Pickles**, *A History of Spaces: Cartographic Reason, Mapping, and the Geo-Coded World* (New York: Routledge, 2004).

21 I discuss this in more length in *Mobile Modernity: Germans, Jews, Trains* (New York: Columbia University Press, 2007), 58ff. The classic cultural and social history of the railway remains **Wolfgang Schivelbusch**, *The Railway Journey* (Berkeley: University of California Press, 1977), who also describes the "annihilation" of space and time.

22 **Walter Benjamin**, "A Berlin Chronicle," in: *Walter Benjamin: Selected Writings, Volume 2: 1927-1934*, eds. Michael Jennings, Howard Eiland, and Gary Smith, trans. Edmund Jephcott (Cambridge: Harvard University Press, 1999), 596.

23 Such a project begins to approach the tripartite analysis of space as lived, perceived, and represented that Lefebvre articulated in *The Production of Space*, trans. Donald Nicholson-Smith (Cambridge: Blackwell, 1991).

24 **Walter Benjamin**, "Berlin Childhood Around 1900," in: *Walter Benjamin: Selected Writings, Volume 3: 1935-1938*, ed. Michael Jennings, trans. Howard Eiland (Cambridge: Harvard University Press, 2002), 387. Translation modified.

25 **Philip Ethington** argues that "knowledge of the past . . . is literally cartographic: a mapping of the places of history indexed to the coordinates of spacetime," in: "Placing the Past: 'Groundwork' for a Spatial Theory of History," *Rethinking History* 11.4 (2007): 465-93. Here, 466.

26 **Jules Michelet**, *Histoire de France* (Paris: C. Marpon and E. Flammarion, 1879-1884), 2.

27 The theoretical claims made in this essay are developed extensively in: **Philip J. Ethington**, "Placing the Past: 'Groundwork' for a Spatial Theory of History," with responses by **Thomas Bender**, **David Carr**, **Edward Casey**, **Edward Dimendberg**, and **Alun Munslow**, *Rethinking History* 11:4 (December 2007): 463-530; Philip J. Ethington, "Sociovisual Perspective: Vision and the Forms of the Human Past," in Barbara Stafford, ed., *A Field Guide to a New Meta-Field: Bridging the Humanities-Neurosciences Divide* (Chicago: University of Chicago Press, 2011); and Philip J. Ethington and **Nobuko Toyosawa**, "Inscribing the Past: Depth as Narrative in Historical Spacetime," in David Bodenhamer, ed., *Deep Mapping and Spatial Narratives* (Bloomington: Indiana University Press, 2014).

28 **Denis Cosgrove**, *Apollo's Eye: A Carto-*

graphic Genealogy of the Earth in the Western Imagination (Baltimore: Johns Hopkins University Press, 2001).

29 See the excellent overview by **Benjamin Lazier**, "Earthrise; or, The Globalization of the World Picture," in: *The American Historical Review* 116.3 (June 2011): 602-30.

30 **Martin Heidegger**, "The Age of the World Picture," in: *The Question Concerning Technology and Other Essays*, trans. William Lovitt (New York: Harper & Row, 1977), 129.

31 Cf. *Media Archaeology: Approaches, Applications, and Implications*, eds. Erkki Huhtamo and Jussi Parikka (Berkeley: University of California Press, 2011).

32 **Susan Sontag**, *Regarding the Pain of Others* (New York: Farrar, Straus, and Giroux, 2003), 110.

33 **Jorge Luis Borges**, "On Exactitude in Science," in: *Collected Fictions*, trans. Andrew Hurley (New York: Penguin, 1999).

34 **Lev Manovich**, *The Language of New Media* (Cambridge: MIT Press, 2001), 260ff. For a discussion of Shaw and Viola, see Mark B.N. Hansen, *New Philosophy for New Media* (Cambridge: MIT Press, 2004).

35 http://www.artcom.de/en/projects/project/detail/the-invisible-shape-of-things-past/.

36 **Stephen Oettermann**, *The Panorama: History of a Mass Medium*, trans. Deborah Lucas Schneider (New York: Zone Books, 1997), 49.

37 **Jonathan Crary**, *Techniques of the Observer: On Vision and Modernity in the Nineteenth Century* (Cambridge: MIT Press, 1990), 1-2. Also, see the discussion by Mark B.N. Hansen in *New Philosophy for New Media*, esp. chapter three, "The Automation of Sight and the Bodily Basis of Vision."

38 It is not coincidental that the German word for the sublime, *Das Erhabene*, comes from the verb *erheben*, meaning to elevate or rise up. With regard to the pleasure of rising above the earth, Cosgrove explains: "Reverie is the closest English translation of the Latin *somnium*, the sense of imaginative dreaming long associated with rising over the earth" (3).

39 **Caren Kaplan**, "Precision Targets: GPS and the Militarization of U.S. Consumer Identity," in *American Quarterly*, 58.3 (September 2006), 693-713. Here, 695.

40 **Lisa Parks**, *Culture in Orbit: Satellites and the Televisual* (Durham: Duke University Press, 2005), 14.

41 For a discussion of the key satellites used in remote sensing from 1960 through the present, see the excellent overview by **Laura Kurgan**, *Close Up at a Distance: Mapping, Technology and Politics* (New York: Zone Books, 2013), 39-54. Her book appeared just as we were finishing this book, and there are many parallels between HyperCities and the nine projects she discusses that make use of geospatial technologies, satellite imagery, and GIS.

42 **Patrick Crogan**, "Gametime: History, Narrative, and Temporality in Combat Flight Simulator 2," in: *The Video Game Theory Reader*, eds. Mark J.P. Wolf and Bernard Perron (New York: Routledge, 2003), 275-301. Here, 276.

43 **Manovich**, *The Language of New Media*, 278.

44 According to Google Earth statistics, most of the imagery is between one and three years old and is periodically updated as higher resolution imagery becomes available. Moreover, quite a number of militarily sensitive regions of the earth, such as the roof of the White House or a host of worldwide military bases and secret gov-

ernment buildings, are deliberately obscured or simply erased, something that, perhaps ironically, only draws more attention to these "empty" spaces.

45 For a discussion of the "unknown world" in the history of cartography, see **Alfred Hiatt**, "Blank Spaces on the Earth," *The Yale Journal of Criticism* (2002): 15.2, 223-250.

46 The Van Sant map cannot be reproduced in print but can be seen at this website, which proclaims it as "a milestone in cartographic history": www.geosphere.com/imagery.html. For a trenchant analysis of this map and its ideology of visual purity, see the discussion by **Denis Wood**, *The Power of Maps* (New York: Guilford Press, 1992), 48-69. See also, the short discussion by **John Pickles**, *A History of Spaces: Cartographic Reason, Mapping, and the Geo-Coded World* (New York: Routledge, 2004), 62.

47 Google Earth shows the city's built environment (*urbs*) in virtual form, not the social and economic structures of the community (*civitas*). For a discussion of this distinction, see **Richard Kagan**, *Urban Images of the Hispanic World, 1493-1793* (New Haven: Yale University Press, 2000).

48 **Max Horkheimer** and **Theodor Adorno**: *Dialectic of Enlightenment*, trans. John Cumming (New York: Continuum, 1993), 3.

49 The panoramic bubbles were introduced in version 5 of Google Earth, but removed in subsequent versions.

50 **Bruce Sterling** calls this an "internet of things," in which data networks have merged with physical objects in space. See his *Shaping Things* (Cambridge: MIT Press, 2005).

51 Established by the inventor of gmail, **Paul Buchheit**, Google's informal corporate motto is "don't be evil."

52 These words were written in his preface in 1988. **Paul Virilio**, *War and Cinema: The Logistics of Perception*. trans. Patrick Camiller (London: Verso 2000), 4.

53 **Avi Bar-Zeev**, "Notes on the Origin of Google Earth": http://www.realityprime.com/articles/notes-on-the-origin-of-google-earth.

54 Keyhole corporate website, accessed via the Internet Archive Wayback Machine, http://web.archive.org/web/20030721011001/http://keyhole.com/.

55 Conversation between **John Hanke** and **David Shepard** over Twitter Direct Message.

56 http://www.gearthhacks.com and http://googlesightseeing.com/.

57 **Mark Aubin**, "Google Earth: From Space to Your Face . . . and Beyond," Google Librarian Center Blog. Although no longer available, the blog is cited by others.

58 **Avi Bar-Zeev** cites another inspiration for Google Earth namely the "Earth" geographic information system in Neal Stephenson's *Snow Crash* that the Central Intelligence Corporation uses to tie sites in the Internet to the physical world (Bar-Zeev).

59 **John Frick** and **Joel Headley**, "Why does Google maps use the inaccurate, ancient and distorted Mercator Projection?—Google Groups," Google Earth Product Forums: http://productforums.google.com/forum/#!topic/maps/A2ygEJ5eG-o, August 3, 2009. Last visited July 26, 2012.

60 http://support.esri.com/en/knowledgebase/GISDictionary/term/georeferencing.

61 **Clifford Geertz**, *The Interpretation of Cultures*, 28-29.

62 **Jerome McGann**, *Radiant Textuality: Literature after the World Wide Web* (New York: Palgrave Macmillan, 2001), 117.

63 The following summary is based on research undertaken by **Nobuko Toyosawa**, which she graciously shared with our research team during a NEH Summer Institute for Advanced Topics in Digital Humanities focused on geospatial humanities scholarship (2012).

64 The projects discussed by **Laura Kurgan** in her book, *Close Up at a Distance*, use geotechnologies and critically explore spatial data and GIS systems—much like HyperCities—in order to "reclaim, repurpose, and discover their inadvertent, sometimes critical, often propositional, uses" (17) and, thereby, "assess their full ethical and political stakes" (14).

65 This is the challenge at the conclusion of **Fredric Jameson**'s *Postmodernism, or the Cultural Logic of Late Capitalism*, 54.

66 **Jean-François Lyotard**, *The Postmodern Condition: A Report on Knowledge*, trans. Geoff Bennington and Brian Massumi (Minneapolis: University of Minnesota Press), 60ff.

67 For an overview of modalities of "collaboration" and "participation" in the digital humanities, see **Lisa Spiro**'s essay, "Computing and Communicating Knowledge: Collaborative Approaches to Digital Humanities Projects": http://ccdigitalpress.org/cad/Ch2_Spiro.pdf.

68 **Michel Foucault**, "The Discourse on Language," in: *The Archaeology of Knowledge*, trans. A. M. Sheridan Smith (New York: Pantheon, 1970).

69 **Jacques Derrida**, "The University without Condition," in: *Without Alibi*, trans. Peggy Kamuf (Stanford: Stanford University Press, 2000), 202-37.

70 Voices of January 25th: https://twitter.com/Jan25voices and Voices of February 17th: https://twitter.com/feb17voices.

71 For a sobering account of the dialectical underbelly of social media technologies, see, for example, **Evgeny Morozov**'s analysis of the "dark side" of the Twitter revolution, in which he shows how totalitarian regimes harness social media to track and detain dissidents: "Iran: Downside to the 'Twitter Revolution,'" *Dissent* (Fall 2009): 56.4, 10-14, and *The Net Delusion: The Dark Side of Internet Freedom* (New York: Public Affairs, 2011).

72 **Paul Virilio**, "The Third Interval: A Critical Transition," in: *Rethinking Technologies*, ed. Verena Conley (Minneapolis: University of Minnesota Press), 9.

73 See the blog post by **Ramesh Srinivasan**: "Starting with Culture Before Technology—My Work From Egypt" (August 30, 2011): http://rameshsrinivasan.org/2011/08/30/starting-with-culture-before-technology-my-work-from-egypt/.

74 **Ramesh Srinivasan**, "Digital Dissent and People's Power: Ramesh Srinivasan at TEDxSanJoaquin," TEDxSanJoaquin, October 10, 2012, San Joaquin, CA.

75 **Gilles Deleuze**, *The Logic of Sense*, trans. Mark Lester (New York: Columbia University Press), 64.

76 **Lev Manovich** on "cultural analytics": http://www.manovich.net/cultural_analytics.pdf.

77 **Franco Moretti**, "Conjectures on World Literature," in: *New Left Review* (Jan.-Feb. 2000): 54-68.

78 **Johanna Drucker**, "Humanities Approaches to Graphical Display," *Digital Humanities Quarterly 5.1* (2011), 2. http://www.digitalhumanities.org/dhq/vol/5/1/000091/000091.html.

79 Since the Revolution, social media access and use has grown throughout Egypt, not only giving rise to new ways to influence and undermine traditional media, but also enfranchising a significantly broader spectrum of activists and politicians—ranging from the Muslim Brotherhood to the Egyptian military. **Ramesh Srinivasan**, "Taking Power Through Technology in the Arab Spring," in *Aljazeera* (October 25, 2012): http://www.aljazeera.com/indepth/opinion/2012/09/2012919115344299848.html.

80 All network graphs were generated using the Gephi network visualization software.

81 To extract the hashtags for the graph, we used the following algorithm to include non-English characters: split the tweet by whitespace into tokens, find any that begin with a hash sign (#), and keep all characters in them before the first punctuation mark (as defined in Python's string.punctuation) or ANSI control characters (non-printing text-display commands) before Unicode character code 32 (0x1F). Gephi seems able to display the Arabic characters, but the monolingualism of some text analysis algorithms is a significant problem that must be addressed by global digital humanities. After all, to truly enable "participation without condition," the technologies must support linguistic diversity and not simply extend the cultural hegemony of English, despite the technical challenges in adapting text-searching algorithms to non-English characters.

82 Louvain was run in Gephi randomized, using edge weights, with a resolution of 1.0. The same settings were used for all subsequent community detection operations.

83 **Derrida**, "Structure, Sign, and Play in the Discourse of the Human Sciences," *Writing and Difference*, trans. Alan Bass (London: Routledge, 2001), 351-70.

84 Roughly speaking, the Louvain method operates by placing every node into a community, and checking that community's connection to other communities. If a community has a strong connection to another community, the method places the node in that community. Then, it applies the same process to the communities it has created, and continues creating communities of communities of communities until it reaches a certain limit. For a complete description of the method, see **Vincent D. Blondel**, **Jean-Loup Guillaume**, **Renaud Lambiotte**, and **Etienne Lefebvre**, "Fast Unfolding of Communities in Large Networks," *Journal of Statistical Mechanics: Theory and Experiment* (October 2008): 2008.10, n.p. (web).

85 **Yoh Kawano**, Hiroyuki Iseki, "GIS Volunteering: Tohoku Kanto Earthquake in Japan." http://www.giscorps.org/index.php?option=com_content&task=view&id=102&Itemid=63.

86 **Olivia Katrandjian, Joohee Cho, Juju Chang,** "Japan's Fukushima 50: Heroes Who Volunteered to Stay Behind at Japan's Crippled Nuclear Plants," *ABC World News* (March 16, 2011): http://abcnews.go.com/International/fukushima-50-line-defense-japanese-nuclear-complex/story?id=13147746#.UKyLlKX3BL8.

87 "Nuclear Workers in Japan: Heroism and Humility," *The Economist* (October 27, 2012): http://www.economist.com/news/asia/21565269-meet-%E2%80%9C-fukushima-50%E2%80%9D-men-front-line-nuclear-disaster-heroism-and-humility.

88 A cron job is a Unix program run at a

scheduled time by the operating system, instead of in response to user input.

89 **Nakamura Akira**, *Emotion Expression Dictionary* (Tokyo-do Shuppan, 1993) [in Japanese].

90 The original methodology to detect emotions through text matching was suggested by **Eiji Aramaki** PhD. Eiji is a natural language expert working for the Center for Knowledge Structuring, and is currently an assistant professor at Tokyo University.

91 The 35 day "text" files were exported as separate CSV files from a MySQL database. These files were first tokenized using MeCab (an open source Japanese language morphological analyzer created by Kyoto University), then imported into WordSmith where multiple concordance routines were run against each group of emotion keywords. The resulting output was transformed into "nodes" and "edges" and subsequently imported into Gephi in order to generate the network diagrams.

92 **William Pesek**, "A Disaster Made in Japan," *Bloomberg* (July 9, 2012): http://www.bloomberg.com/news/2012-07-09/search-for-truth-finds-a-disaster-made-in-japan.html.

Acknowledgments

The HyperCities project—a decade-long effort to explore thick mapping in the digital humanities—was made possible by the support of many people, organizations, and foundations. First, we want to acknowledge the extraordinary creativity and acuity of our technical development team, particularly Chen-Kuei "James" Lee and Lung-Chih "Jay" Tung, both (at the time) graduate students in UCLA's Computer Science department. They worked closely with Presner, Shepard, and Kawano over the last five years to design HyperCities and support projects developed using a wide variety of mapping technologies. Collectively, the dedication and insights of the "technical team" were indispensable for the successes of the project. We also thank the "HyperCities Collaborative," a team of faculty and community partners, convened by Presner, project founder and director: Dean Abernathy, Mike Blockstein, Reanne Estrada, Philip Ethington, Diane Favro, Chris Johanson, John Maciuika, Jan Reiff, and Aquilina Soriano-Versoza. Thanks also go out to the many students who contributed to the project over the years, especially Ryan Chen, Barbara Hui, Albert Kochaphum, Patrick Tran, Andy Trang, Shawn Westbrook, and Elliot Yamamoto. Andy and Patrick created much of the project documentation and, over the years, supported many students in using HyperCities as well as produced several exemplary projects themselves, including a series of mappings of the life stories of local Holocaust survivors. Ryan, Albert, and Elliot were instrumental as GIS assistants and played a central role in creating the maps in the book as well as helping with the book's overall design. We also thank UCLA's Institute for Digital Research and Education (and especially Lisa Snyder, Grant Young, and Christina Patterson), the UCLA Library (Gary Strong, Todd Grappone, Zoe Borovsky, Jonnie Hargis, and Stephen Davidson), and the UCLA Center for Digital Humanities (Shawn Higgins, Yusuf Bhabhrawala, Annelie Rugg, and Akash Haswani). Over the years, countless many students, faculty, and members of the general public have provided valuable feedback and opportunities for collaboration, and we would particularly like to single out: Cathy Davidson, Craig Dietrich, Amir Eshel, David Theo Goldberg, Laura Mandell, Lev Manovich, Tara McPherson, and, most recently, the cohort of the 2012 NEH Summer Institute at UCLA on Advanced Topics in the Digital Humanities. Finally, we want to acknowledge the vision and support of Jeffrey Schnapp, who provided the very first round of seed-funding in 1999 for what would become HyperCities and has been a stalwart supporter of the project ever since. Together with our editor, Michael Fischer, his guidance has been instrumental in producing this book.

Regarding the Japan story, we would like to thank our partners at Niigata University, particularly Makoto Naito, Jun Goto, and Yugo Shobugawa, who continue to work tirelessly toward improving the lives of those affected by the disaster. We would also like to thank Satoshi Mimura who kindly volunteered an entire day to drive Kawano and the team from Niigata University around the disaster areas, and to the government workers from the City of Namie for accompanying us into the nuclear evacuation zone. Finally, the HyperCities project graciously acknowledges the visionary and critical support of the MacArthur Foundation and HASTAC for the 2008 "digital media and learning prize," the Haynes Foundation, the Keck Foundation, the National Endowment for the Humanities, the American Council of Learned Societies, and a Google "Digital Humanities" award.

Credits

Figure 1, page 25: Google Earth
Figure 2, page 27: The Granger Collection, New York
Figure, 8, page 84: NASA
Figure 9, page 87: NASA
Figure 10, page 89: Google Earth
Figure 13, page 101: Google Earth

Figure 16, page 115: Used by permission.

Figure 19, page 117: Used by permission.

Author Key

TP - Todd Presner
DS - David Shepard
YK - Yoh Kawano